THE MODERN SHEPHERD

Also by Sam Meadowcroft

INDOOR BEEF PRODUCTION
(with Ron Hardy)

THE MODERN SHEPHERD

Dave Brown and Sam Meadowcroft

FARMING PRESS

First published 1989

British Library Cataloguing in Publication Data

Brown, Dave
The modern shepherd.
1. Livestock: Sheep. Production
I. Title II. Meadowcroft, Sam
636.3

ISBN 0-85236-188-2

Published by Farming Press Books
4 Friars Courtyard, 30–32 Princes Street
Ipswich IP1 1RJ, United Kingdom

Distributed in North America
by Diamond Farm Enterprises,
Box 537, Alexandria Bay, NY 13607, USA

Typeset by Galleon Photosetting, Ipswich
Printed and bound in Great Britain by Butler and Tanner, Frome, Somerset

Contents

Acknowledgements

The authors have drawn heavily on their observations over the last 30 years of the ewe flock at Rosemaund Experimental Husbandry Farm, Hereford. We are therefore grateful for the help given by colleagues past and present over this period. Thanks are also due to the many individuals and organisations who have provided us with information, figures and photographs included between these covers. In particular, many members of ADAS have offered advice, and both MAFF and MLC publications have been extensively used.

We are grateful to the following ADAS colleagues for reading the book and making many helpful suggestions: Geoff Barnes, Colin Slade and Barbara Maund (Livestock Husbandry), Pete Kelly (Nutrition Chemistry), Bill Hall (Farm Management) and Eddie Winkler (Veterinary Investigation Officer). In the section on marketing we relied heavily on the advice of Alan Barrett of Barrett & Baird Ltd.

The typing was in the hands of Mary Hack, Isle of Rhea Cottage, Bodenham, Hereford, and for her great skill and patience we are very grateful, as we are to our wives Marg Brown and Pat Meadowcroft for their help and support over the last 18 months.

Preface

Most books on the subject of sheep farming make pessimistic reading. Their authors contrast the considerable technical advances in many farm enterprises such as dairying and cereal growing with the pedestrian pace of progress in the sheep enterprise, and are led to question the viability of the sheep flock on the lowland farm.

Support for this view comes from surveys of the performance of UK flocks. These show that many produce financial margins too low even to cover their share of the farm overheads. They are secondary enterprises only justified because the capital requirement is relatively low and the sheep are valued for their beneficial effect on soil fertility.

This book has been written because the authors believe that the current level of performance is abysmally low and that we now have the technical expertise to double the average levels of lamb output and gross margin per hectare. We do not believe that the flockmaster's salvation lies in buying expensive, highly prolific ewes which will drop litters of undersized lambs. It lies in the adoption of the best of the husbandry practices developed in recent years. These are fully explained in the book and it is remarkable that many of them cost very little.

The techniques described ensure high output per ewe and per hectare, and when they are combined with good traditional stockmanship and with careful marketing, the result can be a sheep enterprise more profitable than cereal growing.

May 1989

Dave Brown
Sam Meadowcroft

PART I

The Cinderella Enterprise

Chapter 1

The Sheep Industry

SHEEP IN THE EC COUNTRIES

This short introductory chapter outlines the state of the sheep industry in the EC and in the UK.

There is a huge difference in terms of output and profitability between the average and the best sheep farmers. This chapter shows that sheep farming must become more competitive as we approach full integration within the EC, that the general standard of management of the average flock will therefore have to improve, and also pinpoints those aspects of sheep production which give scope for improvement. Only if opportunities for improvement are seized will the 'Cinderella enterprise' of UK farming cease to be a secondary activity, and become fully competitive with other crop and livestock enterprises.

In Table 1.1, figures are not quoted for Denmark and Luxemburg. The total sheep population of the EC in December 1985 was around 84 million. Sheep numbers are steadily increasing in most countries, including the UK.

Table 1.1 Total sheep populations in the EC countries (millions)

	December 1984	*December 1985*	*December 1986*	*December 1987*
Belgium	0.11	0.13	0.12	0.13
West Germany	1.30	1.30	1.38	1.41
Greece	10.03	9.99	11.03	10.82
France	10.82	10.79	10.58	10.36
Eire	2.69	2.77	2.91	3.25
Italy	11.29	11.29	11.45	11.48
Netherlands	0.92	0.95	1.11	1.21
UK	23.95	24.54	25.97	27.47
Spain	—	—	17.87	20.29
Portugal	—	—	3.00	3.03

Source: EC.

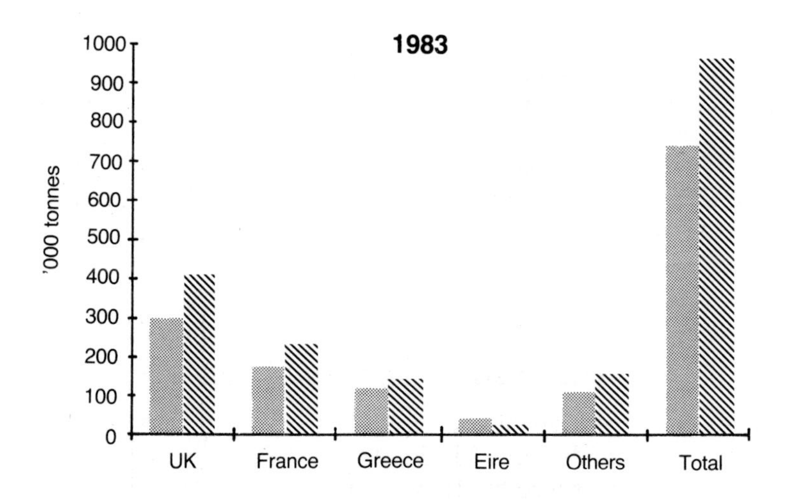

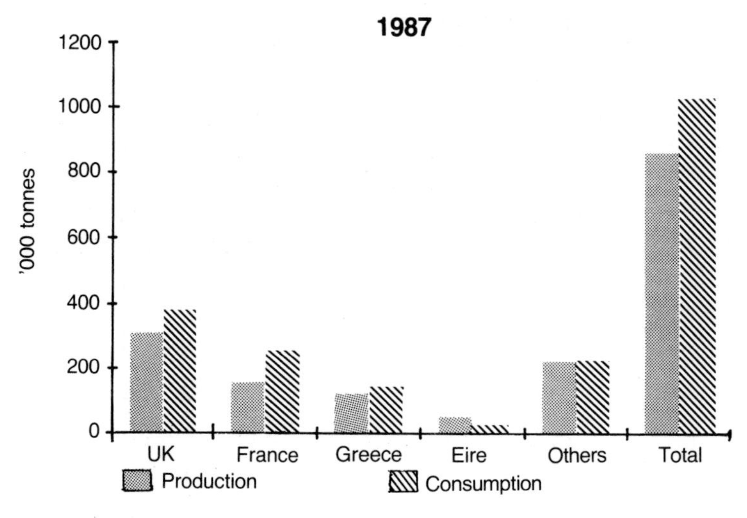

Source: CFR Slade

Figure 1.1 Lamb production and consumption in 1983 and 1987

The levels of production and consumption of sheep meat in the EC in 1983 and in 1987 are shown in Figure 1.1. It can be seen that the UK is the major sheep-meat producer in the EC. In Greece, a substantial number of sheep are milked; hence, the level of meat production is not proportional to the sheep population. In Spain, the present requirement is for small carcasses and here again the contribution to meat production is less than the sheep population might suggest.

With the exceptions of Ireland and Portugal, all EC countries consume more sheep meat than they produce. In 1983 the EC was 77 per cent self-sufficient in sheep meat but this figure had increased to 84 per cent by 1987. Similarly, the UK was 73 per cent self-sufficient in 1983 but by 1987, because of increased ewe numbers, the level of self-sufficiency had increased to 81 per cent. The main outside suppliers of sheep meat to the EC are New Zealand and Eastern European countries.

THE POSITION WITHIN THE UK

The increase in the size of the national flock has been considerable during the last decade, and MAFF figures show that the number of sheep in the UK increased by 45 per cent between 1975 and 1988. Yet during this period of increasing sheep meat production, the amount of lamb and mutton consumed per person in the UK has declined. Consumption figures from MAFF for 1987 show that less lamb is eaten than beef, pork, ham or poultry. This is illustrated in Figure 1.2.

The consumption of lamb has been declining for over 20 years and supermarket surveys suggest that this decline is more marked in customers under 40 years of age. One reason for this is that lamb has not been channelled into the convenience food market. Today's working housewife spends much less time preparing food than her

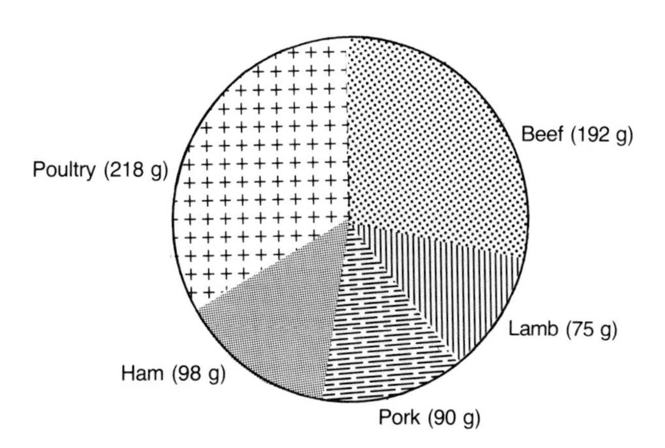

Source: MAFF

Figure 1.2 Types and amounts of meat consumed in the UK in 1987 (person/week)

predecessors and relies a great deal on 'quick meals' for her family. Also, the relatively small muscle size of lamb compared with other meats makes it less attractive. The new styles of cuts pioneered by the MLC are slowly making an impact, and 'lamburgers' are now available in shops. The trade is slowly adapting to meet modern requirements.

The greatest inhibitor to increased lamb consumption is the number of overfat lambs produced. Not only does the fat look and taste unattractive, but today's health-conscious families are aware that too much fat can lead to heart disease.

Although the standard lamb carcass in the UK weighs 15–20 kg, some meat buyers require lambs weighing 23–25 kg for the processing trade. This is in part because it takes just as much time to strip, process and joint a small carcass by hand as it does a large one. Lambs from the majority of breeds and crosses of sheep in the UK tend to become too fat before they reach these carcass weights. Although sheep suitable for the production of large carcasses have been bred, the present thinking is that in the near future many lamb carcasses will be stripped mechanically. The size of the carcass would then matter less.

Table 1.2 The effect of fat class on carcass meat yield

	Fat class						
	1	*2*	*3L*	*3H*	*4L*	*4H*	*5*
Saleable meat (%)	95.0	93.0	92.0	91.0	90.0	89.0	85.5

Source: MLC.

It is apparent from Table 1.2 that the fatter the carcass, the lower the percentage of saleable meat it contains. It is hardly surprising that the trade does not want fat lambs. So although producers may take more money in the short term by selling heavier, fatter lambs, this policy will discourage the housewife from buying lamb, and in the long term, profitability will be lower as the producer becomes more dependent on the market for his returns.

The scope for improvement

The overall productivity of the national flock can only be described as poor. Sheep have justified their presence on the hills because of low land values and a lack of alternative enterprises. In the lowlands their poor performance has been tolerated because of their beneficial effects on soil fertility and because they can be used to graze the ley break in the rotation without a large commitment of capital or labour.

During the 25 years following the end of the Second World War there was little technical innovation and no economic encouragement for the flockmaster. However, the accession of the UK to the EC has encouraged sheep production. The EC subsidies have improved the financial returns from the sheep enterprise. Moreover, since 1970 there have been a number of technological developments which have enabled the flockmaster to run an enterprise which can stand on its own feet in providing a reasonable return on invested capital. The more important of these developments point the way to improvements in the following areas.

- The weight of lambs sold per ewe.
- The weight of lambs sold per hectare.
- The accuracy of the winter feeding of the ewe.
- The ability to market lambs throughout the year.
- Disease prevention.

Weight of lambs sold per ewe

Weight of lambs sold per ewe is the product of the number of lambs reared per ewe and their average sale weight. In spring-lambing lowland flocks the number of lambs reared per 100 ewes put to the tup ranges from 100 to 200. The latest results from MLC recorded flocks (1987) show that in the average flock, 1.51 lambs/ewe put to the ram were reared, with the top-third flocks rearing 1.58. These performances fall well short of the sale of 2 lambs/ewe which is being attained by the very best flockmasters. Correct ewe management during the pretupping and pregnancy periods can improve prolificacy, and accurate feeding and disease prevention techniques will reduce lamb mortality.

Weight of lambs sold per hectare

Weight of lambs sold per ewe × ewe stocking rate per hectare equals weight of lambs sold per hectare. Therefore, the stocking rate has a big influence on the financial return, and here again big improvements can be made. The MLC recorded flocks (1987) show an average stocking rate of only 12.5 ewes/ha, whereas the figure could be nearer the 20 ewes/ha already achieved on some farms in this country and in New Zealand.

Recent experience at Rosemaund Experimental Husbandry Farm (EHF) has convinced the authors that on a good-quality ryegrass/white clover sward at least 15 ewes/ha can be grazed with only modest levels of fertiliser usage. At 1.8 lambs/ewe reared and an average live selling-weight of 36 kg/lamb, this means that 27 lambs/ha would be reared and 972 kg/ha of live lamb would be produced. This stocking rate may be increased further if the need for grass conservation is removed by feeding straw as the only roughage allowed the ewes during the winter period.

Of course, high stocking rates can be combined with good animal performance only if the correct steps are taken to control parasitic worm infections.

Better winter ewe management

The wide adoption of winter housing for in-lamb ewes has undoubtedly provided an opportunity to improve the level of management and the precision of feeding. The advantages are lower lamb mortality and better milk yields leading to higher growth rates in the lambs. Housing the flock during the winter also allows higher year-round stocking rates on most farms.

Year-round marketing of lambs

There is clearly a need to even out further the pattern of lamb marketing through the year (Table 1.3). Increased lamb supplies from January to June may be achieved in two ways, firstly by out-of-season breeding techniques and, secondly, by finishing more hoggets from the New Year onwards.

In Part II, recent technological advances which have proven advantages for the flockmaster are discussed. Part III shows how they can be integrated with the traditional skills of the shepherd. Provided that attention to detail is observed in the management of the flock, and that the quality of lamb which we offer for sale is right for the market, the sheep industry can look optimistically to the future.

Table 1.3 Slaughterings by season in the UK

	1981 (%)	*1987* (%)
Jan.–March	19	22
April–June	16	18
July–Sept.	36	30
Oct.–Dec.	29	30

Source: CFR Slade.

PART II

Recent Husbandry Developments

Chapter 2

Increasing the Lamb Crop

NUMBERS OF LAMBS AND LAMB OUTPUT

The title of this chapter was chosen with some care because the number and weight of lambs sold per ewe put to the tup is a key factor in flock profitability, and the latter is the main measure of efficiency in which the bank manager will be interested.

- Number of lambs sold/ewe × average sale price = lamb output £/ewe.
- Lamb output £/ewe × stocking rate in ewes per hectare = lamb output £/ha.

High lamb output per hectare (relative to comparable flocks) combined with reasonably low costs is a sure indication of good management. Sales of wool and of culled ewes are usually relatively small outputs.

Several other terms which are used to describe reproductive efficiency (e.g. litter size, lambing percentage and prolificacy) should be regarded with suspicion.

Litter size refers to the average number of lambs born to each lambing ewe; it is of interest in giving some indication of the proportions of single and multiple births, but as an indicator of flock performance it can be extremely misleading. This is well illustrated in Table 2.1, from a paper written by Barbara Maund (ADAS) and Richard Jones (NAC sheep unit manager), in which it can be seen that with an identical litter size the lambing percentage in year 2 was well up on that in year 1.

Lambing percentage (prolificacy) as given in Table 2.1 gives a good record of breeding performance since it takes into account the ewes which have died, aborted or proved barren as from the time of tupping. However, flockmasters who calculate lambing percentage per 100 ewes lambed-down and not per 100 ewes put to the tup are deluding themselves (and others) by 'forgetting' the ewes which were non-producers in that year.

Even 'honest' lambing percentage records are not good measures

Table 2.1 Flock lambing performance in 2 years

	Year 1	*Year 2*
Ewes to the ram	100	100
Ewes died	3	1
Ewes culled	3	—
Ewes aborted	8	1
Ewes barren	3	2
Ewes lambed	83	96
Lambs born	166	192
Mean litter size	2	2
Lambing %	166	192

Source: NAC *Sheep Unit Newsletter*, December 1986.

of flock performance unless account is taken of the number of lambs born dead and those lost before weaning and between weaning and sale. Only then do we reach the true yardstick of the size of the lamb crop: the number of lambs sold per 100 ewes to the tup.

An example is the histogram in Figure 2.1 showing the 1987 performance of the Rosemaund EHF lowland ewe flock and that of the average and top-third lowland flocks recorded by the MLC and reported in 1987.

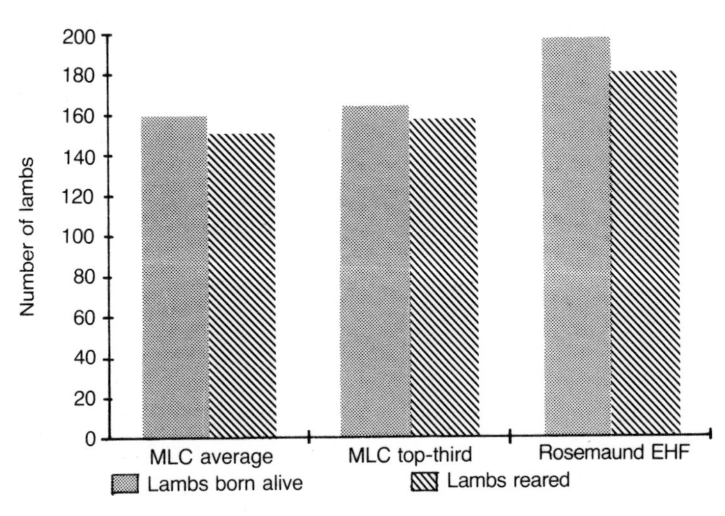

Source: MLC and Rosemaund EHF

Figure 2.1 Numbers of lambs born and reared per 100 ewes

The importance of the number of lambs reared

In the MLC *Sheep Year Book 1988*, the 1987 records of 414 spring-lambing flocks are divided into the top-third performers and the rest. These best flocks out-performed the remainder by £220 in terms of gross margin per hectare, and the MLC analysed the reasons for this superior performance (Table 2.2).

On all known evidence, the number of lambs reared is second only to stocking rate in its effect on flock profitability. All increases in the lamb crop within the range 1.5–2.5 lambs/ewe are likely to increase the

Table 2.2 Contribution to top-third superiority in gross margin per hectare for lowland spring-lambing flocks

	Contribution (%)
Stocking rate	49
Number of lambs reared per ewe	19
Lamb sale price per head	8
Flock replacement cost	13
Feed and forage cost	4
Other factors	7

Source: MLC *Sheep Year Book 1988*.

financial output per ewe, and this is true despite the slower finishing of twins compared to singles. This is not to say that increasing lambing percentage to the full potential of the flock is always valid. More triplets and quadruplets inevitably reduce birth-weights and increase mortality and may be an embarrassment in requiring extra artificial rearing of lambs which means a high input of skilled labour. However, in view of our improved techniques of triplet rearing (Chapter 11) and the importance of lamb output noted above, today's target for the lowland flockmaster should be 1.8 lambs sold/ewe put to the tup.

The achievement of 180 per cent lamb sales

High lamb sales per ewe per year can result from a heavy annual lamb crop, and there are several ways of achieving this. These include a change to a more prolific ewe type, improvement of the existing flock by rigorous selection of breeding stock and the use of certain beneficial husbandry techniques. They can also result from frequent lambing systems. These involve out-of-season lambing at a frequency greater than once each year, and are described in Chapter 7.

Breed Type of Ewe

The hill farmer and the lowland sheep farmer are dependent on one another for their well-being, as many of the ewes used in the lowlands are purchased from farms in the hills.

The desirable characteristics of the hill breeds are their hardiness and ability to thrive and produce and rear a lamb in a hostile environment. Under hill conditions, high prolificacy is not required. To introduce this attribute to provide prolific ewes suitable for the lowlands, hill ewes are mated to a ram of a prolific breed. In the past, the Border Leicester has been the most popular crossing ram, but recently the Bluefaced (Hexham) Leicester has overtaken it. Both are widely used on hill breeds and examples of these crosses are the following.

- Bluefaced Leicester × Swaledale = North Country Mule.
- Bluefaced Leicester × Beulah or
  Specklefaced Welsh Mountain = Welsh Mule.
- Border Leicester × Scottish Blackface = Greyface.

The progeny of these crosses between the hill ewe and the longwool sire, although prolific, do not have the characteristics which make a good 'table lamb'. They are usually light on the hind legs and are slow-growing. The wether lambs do not command premium prices at the abattoir, but this is compensated for by the high prices realised by the ewe lambs sold as replacement ewes for lowland flocks.

To improve carcass quality, a Down breed is used to mate with the crossbred lowland ewes, to produce the 'table lamb'. These Down-breed rams are referred to as terminal sires and are discussed in Chapter 9.

Much has been said, and much more will be said, about the virtues of one crossbred ewe against another. Each breed and each cross has its virtues and its weaknesses.

The North Country Mule is the most popular crossbred ewe in the lowland flock at present. It is available in large numbers, is highly prolific and is a notably good mother. Results from the Rosemaund EHF flock over many years and from numerous commercial flocks show conclusively that this ewe is capable of rearing a 200 per cent lamb crop. Many flockmasters are (quite rightly) satisfied with this level of prolificacy, and feel that they need look no further. Because the Mule is so popular, it commands a good price as a ewe lamb or as a shearling, which means that its depreciation costs are high.

The North Country Mule can be highly recommended, but alternative crossbred ewes such as the Welsh Mule, Greyface and Masham may be cheaper to purchase and can, in the end, give a similar financial margin.

Recently rams of continental breeds have been used on the hill ewes to try to improve the carcass conformation of both the wether lambs sold for slaughter and the ewe lambs sold for breeding. Currently, the Bleu du Maine is being compared with the Bluefaced Leicester as a crossing sire. Oldenburg, East Friesland and Texel rams are also being evaluated. Finnish Landrace rams have been used to breed more prolific lowland ewes, but to the detriment of carcass conformation.

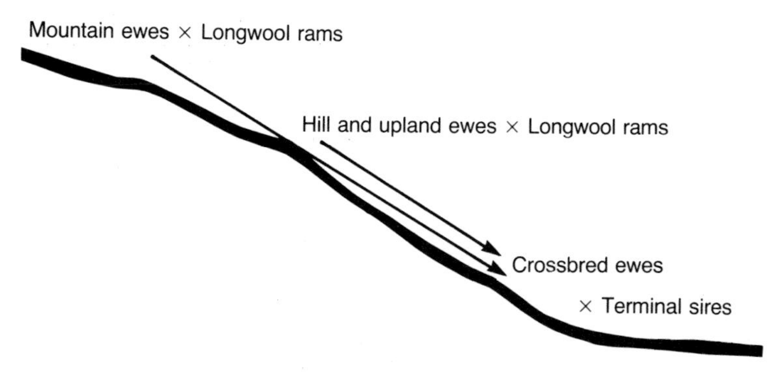

Figure 2.2 Stratification: the reliance of lowland flocks on hill breeds

Pure breeds

Whenever replacement ewes are bought, there is a risk of importing disease; this makes a self-contained flock an attractive proposition. Such a flock of Clun Forest, Kerry Hill, Cheviot or Lleyn ewes may be the answer on some farms. Alternatively, the Cambridge ewe, developed by Professor Owen, has proved very prolific in many commercial flocks. In the Cambridge flock at Drayton EHF, litter sizes of 2.4 were normal. However, this can have its disadvantages: as lambing rises to over 200 per cent, the management has to be very good to take advantage of the full potential.

Husbandry Improvements

Although an increased lamb crop may result from a change in breed type such a change can, of course, be expensive. Moreover, the same objective may be achieved in almost every flock by improving several key aspects of flock husbandry. These improvements can be brought about by simple changes in management and the adoption of known proven techniques. Many of these cost little or nothing, but they are effective. This is why Chapters 8–12 of this book are given over to the detailed description of husbandry techniques designed to increase the lamb crop by optimising the conception rate, the number of lambs born and lamb survival.

Conception rate
A high conception rate is achieved by the best management from weaning, through tupping time and to early pregnancy in the following sequence.

- Wean at the correct time to allow the ewe to recover in condition before mating (Chapter 8).
- Condition-score all ewes following the completion of drying-off. Then feed differentially according to condition (Chapter 8).
- The preparation of rams for tupping time is important (Chapter 9).
- Paying attention to tupping management (Chapter 9).
- Raddling of the tups indicates the speed at which the flock is being served (Chapter 9).

Number of lambs born
The number of lambs born is optimised by the accurate management of the pregnant ewe as follows.

- Ewe feeding throughout pregnancy must match the ewes' needs (Chapter 10).
- Accurate feeding of in-lamb ewes is aided by in-wintering (Chapters 3 and 10).
- It is also promoted by pregnancy scanning of the flock (Chapter 10).
- The worm drenching of in-wintered ewes at housing can increase the lambing percentage (Chapter 10).
- Discuss the avoidance of abortions with your veterinary surgeon and consider vaccination against enzootic abortion (Chapter 11).

Lamb survival
Lamb survival is promoted by reducing the number of weak lambs at birth and minimising losses of new-born lambs:

- The winter shearing of ewes is recommended in the right circumstances. It increases lamb birth-weights and can reduce lamb mortality (Chapter 10).
- All ewes should be vaccinated against the clostridial diseases in order to protect their lambs (Chapter 11).

- High losses of young lambs are not inevitable. Losses may be minimised by providing plenty of skilled labour at lambing time, by taking the correct measures against starvation and hypothermia and by using the best fostering techniques.

Hormonal techniques

It must be remembered that the use of progesterone sponges and of melatonin are in no way a substitute for the best husbandry methods, although they can increase the lamb crop if used in the right circumstances. All these treatments increase the production costs and are therefore most appropriate for high-input – high-output systems such as early lambing and frequent lambing. Ewes on such systems may require help to cycle and ovulate out of season. The use for this purpose of the progesterone sponge and pregnant mare serum gonadotrophin (PMSG) is fully discussed in Chapter 7, as is the use of melatonin. Fecundin, however, has its virtue in increasing the lamb crop, and its use is therefore outlined in this chapter.

Fecundin

Just as it is possible to vaccinate ewes against clostridial diseases, so we can vaccinate the ewes against a hormone which they produce quite naturally, called androstenedione, which has the effect of reducing the number of eggs shed at each heat period. The vaccine (Fecundin) reduces the effects of the hormone, thus allowing more eggs to be shed.

In the first year of using the vaccine, two injections are required, at 8 and 4 weeks before tupping. In subsequent years only a single booster injection given 4 weeks before tupping is necessary.

In extensive trials involving over 2000 ewes, conducted by ADAS in conjunction with Coopers Animal Health, the results shown in Table 2.3 were obtained. Fecundin did not increase the proportion of ewes which lambed, but it did result in extra lambs being born per ewe. The increase in the number of lambs born dead following the use of Fecundin is associated with the increase in the number of multiple births, as shown in Table 2.4.

Fecundin for early lambing

Normally, a lower lambing percentage is to be expected in an early lambing flock than in a flock lambing in the spring, and it was suggested that Fecundin might complement progesterone and PMSG in producing more early born lambs. However, in the few trials which have taken place, the use of Fecundin has not been found advantageous

Table 2.3 Effect of Fecundin on lambing performance

Ewe data	*Control*	*Vaccinated*
Number of ewes	1070	1070
Ewes dead before lambing (%)	1.6	1.3
Ewes barren (%)	3.9	3.5
Ewes aborted (%)	0.3	0.5
Ewes lambed (%)	94.2	94.7
Lambs born/100 ewes mated	153.0	176.0
Lambs born alive/100 ewes mated	149.0	168.0
Lambs born dead/100 ewes mated	4.0	8.0

Source: ADAS.

Table 2.4 Effect of Fecundin on litter size

Litter size	*Frequency of litter size*	
	Control (%)	*Vaccinated* (%)
Single	43.3	29.3
Twin	54.1	59.4
Triplet	2.4	11.1
Quad	0.1	0.2

Source: ADAS.

in the early lambing flock. It has proved useful at Liscombe EHF in a flock tupped late (December) for lambing in May.

Fecundin pros and cons

If the lambing percentage is over 180 per cent, then the extra lambs produced by using Fecundin might well put an unwelcome strain on the farm resources at lambing time. Conversely, if lambing is below 120 per cent, improvement in flock management is more likely to have a beneficial effect than the use of Fecundin.

At present costs it is necessary to sell an extra 12–13 lambs/100 ewes treated to pay for the vaccination. Trials have shown that Fecundin will increase the number of lambs born by 20–30 per cent, but there is no guarantee that all these lambs will be reared and sold.

There is no doubt that the use of Fecundin can increase the lamb crop. In ADAS trials the overall average was increased, although there was a considerable variation in the response. Each flockmaster must decide whether the extra lambs would be reared economically. It must always be remembered that Fecundin is not a substitute for good management and that its use requires extra attention to detail in order to obtain the full benefit of this technique.

Chapter 3

Sheep Housing and Handling

WINTER GRAZING

In the UK most ewes are wintered out of doors. This is the natural management for an animal which is hardy and which has a thick fleece for weather protection. It is also cheaper than putting a roof over its head.

Winter grazing up to the turn of the year may have little effect on the following year's grass production. However, grazing during the first 3 months of the year can damage the swards and thereby limit the flock size to the winter carrying capacity. Where sheep are the main grass-using enterprise, many flockmasters have been forced to consider alternatives to allowing the flock to graze around the farm's grassland throughout the winter. In addition to away-wintering (favoured mainly for ewe lambs), the following possibilities may be considered.

Grazing on forage crops

A range of root and forage crops can be grown to provide sheep grazing throughout the winter. The grazing of crops in situ often results in high feed wastage, variable feed intake and dirty fleeces. In addition, damage is caused to the soil structure by animals and machinery on all but the lightest soils in low-rainfall areas. This also precludes the growing of winter cereal crops, which normally show relatively high financial margins.

Heavy stocking of a sacrifice area

Such an area of grass can be grazed by ewes stocked at up to 35/ha, and the system has been developed by the East of Scotland College of Agriculture and also by Liscombe EHF. However, free-draining land is again essential, shelter is advisable and the inevitable soil structure damage must be accepted.

Winter Housing

Open yards

Cheaper than roofed houses, open yards have been successful in low-rainfall areas, but good floor drainage and draught protection are needed, and the bedding requirement is high.

Roofed housing

Although ewes may be housed for the lambing period only, there is much to be said for housing the intensively stocked spring-lambing flock from late December or early January until 2 days after lambing, a housed period of approximately 12 weeks.

The lowland ewe flock at Rosemaund EHF was first housed in 1963. At the time the majority opinion held that this was crazy. We were told that the provision of sheep housing meant spending money to no advantage, and that the sheep would lose their natural hardiness, if indeed they survived the pneumonia to which many would inevitably succumb.

Certainly, housing does involve an increase in flock costs, but there are considerable compensating advantages. The fears about flock health have proved to be groundless and pneumonia is rarely seen in well-ventilated houses. Indeed, with well-designed buildings and careful management, housed ewes are likely to be healthier than those out-wintered.

Housing: The Advantages

The availability of winter housing gives clear advantages to grassland productivity, flock output and the shepherd's comfort and efficiency.

Grassland productivity

Poaching of grassland caused by winter grazing can, of course, have a drastic effect on its future productivity. But even in the absence of noticeable poaching, experiments have shown that grass growth may be reduced by 15–20 per cent in the following season if the swards are grazed in the January–March period.

In trials at Rosemaund EHF conducted from 1962 to 1964, the effect of winter grazing at a low stocking rate of 5 ewes/ha was to reduce the following season's growth by 15 per cent. Similar results have been reported from other ADAS trials and in those carried out at Cockle Park (Newcastle University Farms) in 1964/65. Here, there was a 20 per cent reduction in grass growth after winter grazing compared to that of grassland rested from the turn of the year.

Perhaps an even more serious consequence of winter grazing is the lack of an early bite for newly lambed ewes in the following March or April. This results in a poorer ewe lactation and subsequent slower growth of lambs which, being weaker, are more likely to succumb to internal parasites. Good grass growth in the spring is the basis for high lamb growth-rates.

Flock output

Higher flock output is the main justification for housing. It can arise from a combination of higher stocking rate, more lambs reared and sold per ewe and an above-average lamb sale price.

Most flockmasters who have housed their ewes have reported an increased stocking rate because the limiting factor of availability of winter grazing has been removed. Where cereal straw is the main bulk feed for housed ewes (Chapter 4), the increase in stocking density is particularly dramatic, reaching 20 ewes/ha or more compared to the traditional 7–8 ewes/ha of the out-wintered flock.

In many flocks, in-wintering has resulted in an increase in the number of lambs born per ewe. Unless this is due to a decision to change to a more prolific ewe type (such ewes are easier to manage indoors in late pregnancy and at lambing), it is difficult to explain. We know that the lambing percentage is largely preordained by events prior to, and at, tupping and in early pregnancy, well in advance of the normal housing date. It can only be assumed that either there is less stress and therefore less foetal wastage in the housed ewe, or that the better spring and summer grass supply promoted by in-wintering brings the ewes to tupping in better condition.

Whether or not more lambs are born to the housed ewe, there is no doubt that many more survive. Over 50 per cent of lamb deaths in the out-wintered flock occur at or soon after birth, and these losses may be considerably reduced in housed conditions. The feeding of the pregnant ewe is more accurate indoors. Ewes can be 'batched' in groups according to their nutritional needs. Young ewes (gimmers and ewe lambs) and the elderly ewes may be penned separately. The remainder of the flock may be penned and fed according to expected lambing date if the tups' raddle colours were regularly changed during the mating period. This is a big management 'plus' in late pregnancy, and results in more uniform ewe condition and lamb birth-weights, and therefore better lamb survival.

At lambing, the better supervision possible indoors, with facilities such as shelter, light, electricity and lambing pens to hand, means easier fostering and fewer lamb losses. Predators such as foxes, carrion crows and rustlers are at a disadvantage! Estimates of the improvement in the number of lambs sold per 100 ewes as a result of housing vary between 5 and 15 per cent.

The housing of ewes facilitates the option of early lambing in December or January (see Chapter 7). Lambs born at this time and sold around Easter realise relatively high prices and further increase the flock output if an acceptable lambing percentage can be achieved.

Shepherd comfort

There is no reason why a shepherd should be expected to work outside through the worst of the British winter weather while a tractor driver is provided with a weatherproof cab. Sheep housing allows him to work in comfort with all facilities to hand, and therefore he can look after more sheep. Some shepherds now tend 800 or more housed ewes with, of course, extra help at lambing and during other operations such as dipping and shearing. They enjoy pleasant working conditions close to home, and housed sheep are quieter and easier to handle. It is noticeable that the low lamb losses in the housed flock are good for the shepherd's morale.

Alternative uses of sheep housing

Most sheep houses have alternative uses during the 9 months out of 12 when not occupied by the spring-lambing flock, and these uses may help to justify their capital cost. Some are used for a flock lambed early in December or January and again for the March lambers. Store lambs may be finished in the house and it can be useful for sheep shearing and storing such things as hay, straw, fertilisers and machinery. The more expensive portal-frame type of house may double as a calf-rearing shed or beef-house.

Housing: The Disadvantages

There are only two significant disadvantages of sheep housing: capital cost and the increased costs of feed and bedding. Lower fleece value is a further disadvantage, but of little importance. The slightly lower lamb birth weights associated with in-wintering in its early days have been eliminated by better feeding of housed ewes. Poorer flock health is frequently ascribed to housed ewes but the authors see no insurmountable problems here if attention is paid to house design, hygiene and preventative medicine. The subject is discussed later in this chapter and in Chapter 10.

Capital cost

This is the one factor which has prevented an even more rapid adoption of the in-wintering technique. It must be remembered that if a new sheep house is to be built, finance will be needed for the house itself, for site preparation and for the provision of electricity and water. Approach roads may have to be constructed and extra feed storage provided, and there will be the cost of internal fitments including pen divisions, feed troughs and water bowls. The size and implications of this expenditure are discussed later in the chapter.

Extra feed and bedding costs

The out-wintered flock obtains some of its feed requirements in the form of grass forage but the housed flock must have this replaced by conserved bulk feeds. At Rosemaund EHF, ewes eat 150 kg of hay or 400 kg of grass silage/head during their housed period of around 90 days. Ewes at grass have a variable need for supplementary hay feeding depending on stocking rate, grass availability and the severity of the winter, but in many winters will require over 50 kg of hay each.

Bedding is an additional requirement of housed ewes. At Rosemaund EHF the requirement has been 3.75 t wheat straw/100 ewes when silage was the main feed, but with straw or hay feeding the bedding requirement was considerably less.

Lower fleece value

It has been a common experience that the clip from housed ewes has been around 0.25 kg/ewe less than from out-wintered sheep, and some flockmasters have reported a slightly poorer fleece quality. It may be that rubbing on pen fitments is responsible for some of the poorer fleeces reported, and a rapid change of diet at housing can cause weaknesses in the fibre.

THE HEALTH OF THE HOUSED EWE

This book is in no way a veterinary text, as the authors are agriculturalists, unqualified to write in detail about sheep diseases. In any case, the reader is recommended to study the number of excellent books on the subject.

However, the flockmaster must anticipate health problems, and should draw up, on paper, a flock health programme. This he must do in close co-operation with his veterinary surgeon in the light of the particular circumstances and history of his farm and flock. The programme in force at Rosemaund EHF is given as an example in Appendix B. Several leading flockmasters have commented to the authors that the correct timing of these health protection measures is all-important.

The veterinary surgeon should not be regarded as someone to be called in only to diagnose a disease already present. We recommend that he should be invited to inspect the flock several times each year, preferably at the critical periods before tupping, during the housed period and at lambing time. After visiting the flock and seeing the facilities and, of course, the sheep, he can suggest any changes necessary to prevent health problems before they arise.

Housed sheep are packed in close together in a semi-enclosed environment for perhaps a quarter of the year, and are therefore inevitably at risk from certain diseases and from their rapid spread. An example is the skin disease orf, which causes painful lesions and which can be transmitted to those handling the affected sheep.

However, reports from many units make it clear that housed sheep are likely to enjoy good health year after year if the following five preconditions are met.

- *Good ventilation* above sheep height with no draughts and a temperature similar to that outside the house. If ventilation is poor, lung disease is a great hazard, and intra-uterine lamb growth may be reduced in overheated ewes.
- *A dry bed*, which can be maintained where there is no condensation and clean bedding is regularly supplied. Damp bedding can result in a serious foot rot problem.
- *Adequate feeding* with no dietary deficiencies.
- *Good hygiene* – especially at lambing time.
- *Careful stockmanship* with the attention to detail which is made easier by keeping sheep indoors.

Please note that extra exercise (achieved by letting the ewes out to grass during the day) is unnecessary. Pregnant ewes do need exercise, but in fact they get this within the house. Housed ewes at Rosemaund EHF were fitted with pedometers, and it was discovered that they were walking about 2 miles each day!

Sheep House Design

The overall objective is to provide cheap shelter for the ewe flock during the winter months. However, the design must satisfy certain basic principles if it is to safeguard the health of the sheep and also satisfy the flockmaster's requirements.

The needs of the sheep

These are detailed in the MAFF Code No. 5 *Code of Recommendations for the Welfare of Livestock: Sheep* as follows.

- The building must have good natural ventilation. This means continuous air movement above the height of the animals but no draughts at floor level. It also has the effect of reducing condensation.
- The building must provide a dry lying area of approximately 1.1 m^2/ewe of average size to allow adequate exercise. This should ideally be divisible into pens holding 30 ewes each.
- The trough frontage must be not less than 450 mm/ewe if rationed concentrates are to be fed. The feeding of hay or silage ad lib requires a frontage of 150 mm/ewe. Water must be available at all times, and small valve-controlled water bowls are preferred.

The needs of the flockmaster

- The site must be well drained and convenient for the provision of a hard approach road, water and electricity. Proximity to the shepherd's house is desirable.
- The house must facilitate a low labour requirement for moving ewes and lambs in and out, for cleaning and for tending the flock, including lambing and feeding. The dimensions of feed passages must be adequate to allow the chosen method of mechanical feeding.
- Whatever the design, the building must be cost-effective. To this end, simplicity of design is desirable, particularly where erection is by farm labour. The provision of concrete floors (except for feed passages and lambing pens) is an unnecessary luxury. Slatted floors

are good for sheep's feet but expensive compared to bedded earth floors.

Types of sheep house

Roofless enclosures cannot be recommended in the wetter areas of the country because of their high bedding costs. However, some flockmasters have reported favourably on part-roofed buildings with a bedded area under cover and a feeding area uncovered. Plastic-covered houses are relatively cheap and are successful in sheltered positions, but the plastic will require regular replacement.

Portal frame steel buildings, although expensive, have a long life and can be put to many alternative purposes. Portable sheep pens may be erected and dismantled within a part or the whole of the building according to need.

Farm-built pole barn sheep houses are extremely cost-effective because of low material (often secondhand demolition timber) and erection costs. However, they are unsuitable for most alternative uses except storage. This type of house has been in use at Rosemaund EHF for 25 years and can be highly recommended.

A simple combined silage trough and pen front was also developed at Rosemaund.

Many and varied sheep houses are on the market, and further advice on design is obtainable from the following organisations.

- ADAS at offices of MAFF.
- National Agricultural Centre, Kenilworth, Warwickshire.

Plate 3.1 Plastic tunnels make cheap sheep houses

Plate 3.2 Good ventilation offered by mono-pitch roof sheep house

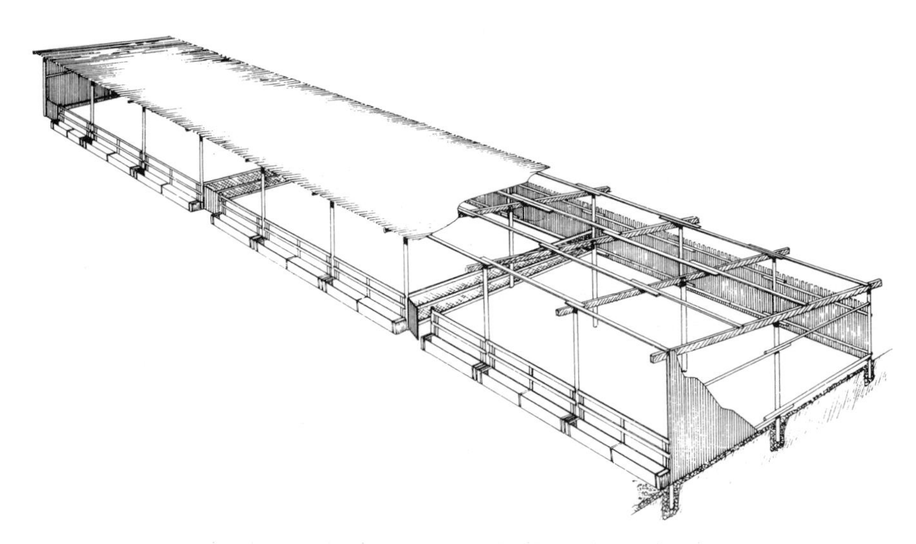

Figure 3.1 General view of a three-pen sheep house for 180 ewes at Rosemaund EHF

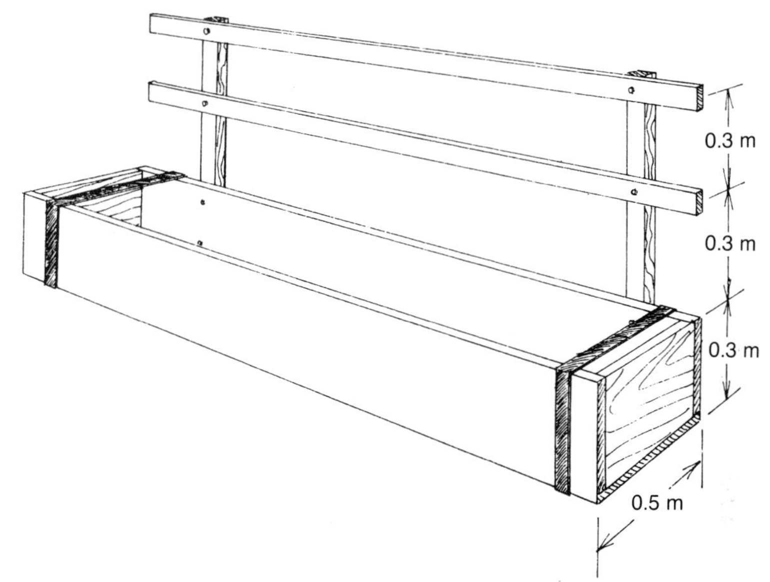

Figure 3.2 Rosemaund combined silage trough and pen front

- North of Scotland College of Agriculture, Craibstone, Bucksburn, Aberdeen.
- West of Scotland College of Agriculture, Auchincruive, Ayr.
- East of Scotland College of Agriculture, Bush Estate, Penicuik, Midlothian.

Does Winter Housing Pay?

The answer is: Only if the annual housing cost per ewe is balanced by improvements in flock output which result from in-wintering.

The annual housing cost is the sum of the annual building charge and other additional costs inherent in the housing of sheep. The first step in the calculation is to work out the annual building charge by ascertaining the capital cost of the house per ewe and amortising this over an appropriate number of years at the current rate of interest. The cost of all types of buildings can vary quite widely, but as a guide, the approximate capital costs of sheep houses in 1988 are as given in Tables 3.1 and 3.2.

The second step is to calculate the annual housing cost per ewe. In the example in Table 3.3, we have used a homebuilt pole barn costing £20/ewe, depreciated over 10 years.

Table 3.1 Cost of structure without fittings

	(£/ewe)
Plastic tunnel	10
Open pole barn (home-built)	20
Portal frame steel building	40

Table 3.2 Calculation of annual building charge (£/ewe at 15 per cent interest rate)

Capital cost of building £/ewe	*Write-off period* 5 years	10 years	15 years
10	3.00	2.00	1.70
20	6.00	4.00	3.40
30	9.00	6.00	5.10
40	12.00	8.00	6.80
50	15.00	10.00	8.50
60	18.00	12.00	10.20
70	21.00	14.00	11.90
80	24.00	16.00	13.60

Table 3.3 Calculation of annual housing cost

	(£/ewe)
Annual building charge	4.00
Repair and maintenance (2% of building cost)	0.40
Straw (40 kg/ewe @ £20/t)	0.80
Additional feed (40 kg hay @ £50/t)	2.00
Annual cost per ewe	7.20

The annual cost per ewe must be balanced by the value of the advantages achieved by housing the flock. Otherwise, winter housing will mean a monetary loss. Most flockmasters have reported that in-wintering has resulted in more lambs being sold and a higher stocking rate at grass, and some have noted an increase in grass conservation yield.

Additional lamb sales

Better lamb survival as a result of housing can increase lamb sales by 5 per cent or more. A 5 per cent increase in lamb sales from 1.60 to

1.68/ewe means, from a flock of 100 ewes, sales increased from 160 to 168.

From 100 ewes 8 extra lambs sold × £38 = £304.00
Value of extra lambs per ewe = £3.04

Increased stocking rate

Since the availability of winter grazing is frequently the factor limiting the stocking rate of the out-wintered flock, the removal of this limitation by in-wintering may allow the stocking rate to rise by 10 per cent.

Flock of 100 ewes × gross margin (GM)/ewe of £35
gives total margin of £3500
Increased flock of 110 ewes × GM/ewe of £35
gives total margin of £3850
Increased margin = £350
Increased margin/ewe = £3.50

Increased conservation

In Rosemaund EHF trials, the absence of winter grazing increased the hay yield in the following season by 625 kg/ha.

Value of 625 kg hay @ £50/t = £31.25
At stocking rate of 15 ewes/ha value/ewe = £2.08

In Table 3.4, the value of the increased output exceeds the additional costs incurred in housing. However, many flockmasters justify the housing of the flock on grounds other than financial. Firstly, they point to the contentment of the shepherd, and the increasing reluctance of the new generation to tend a large out-wintered flock. Secondly, they value the opportunity to improve the management of the flock in the more controlled conditions within the sheep house.

Table 3.4 Sum of value of increased output

	(£/ewe)
Additional lambs	3.04
Increased stocking rate	3.50
Extra hay	2.08
Total	8.62

The winter housing of ewes cannot be recommended on all farms. However, where the land poaches, or the stocking level is high, or the flock lambs early, housing may be considered a near-essential. Once having housed their ewes, few flockmasters return to out-wintering, and a comment frequently heard is: 'We wish we had done it years ago'.

The Sheep-Handling Unit

Every sheep enterprise (large or small) needs a well-designed handling unit. This enables husbandry and veterinary operations to be carried through in a timely manner, which is vitally important. It also ensures efficiency of labour use, so that other farm activities are not hindered because of a high labour requirement in the sheep unit. A good handling unit allows the flock to be held together, sorted and treated with minimum labour input and minimum stress to the sheep and shepherd.

The requirements

The unit must provide suitable facilities for the following operations.

- Dipping against scab (compulsory at present), keds, ticks, lice and blowfly.
- Drenching to control worms.
- Vaccination to prevent disease.
- Foot bathing and trimming to prevent lameness.
- Dagging and veterinary treatments.
- Condition-scoring.
- Marking, weighing, sorting, drafting and loading on to transport.

The site

The site must be chosen with care, be central to the grazing areas and preferably be adjacent to sheep houses. It *must* be well drained and clear of all watercourses because of the danger of pollution by dipping fluids. It should have a good hard access road, water supply and shelter. Large farms with outlying fields may require a central permanent unit close to the housing, and one or more additional portable units servicing blocks of grazing land. On smaller farms, within a ring fence, just one centrally placed handling unit to which the flock can be walked may be sufficient.

The design

It is most important to get the right design at the first attempt, since modifications made at a later date will be expensive. The layout should be simple, but big enough to handle the whole flock at one time (and don't forget possible expansion of flock numbers) including recirculation through the unit.

Sheep behaviour must be taken into account if a smooth flow of animals through the unit is to be achieved. In particular, bear in mind that sheep will follow one another, preferably slightly uphill (an incline of 1 : 60 is recommended), and towards an open prospect. They like to be able to see where they are going, and move most readily away from buildings and people.

Basic to all handling unit designs are the following.

- Collection pens.
- Handling or treatment pen (may be one of the collection pens).
- Drafting race with shedding gates.
- Foot bath (may be in the drafting race).
- Dipping tank.

Plate 3.3 Simple sheep handling unit at Rosemaund EHF

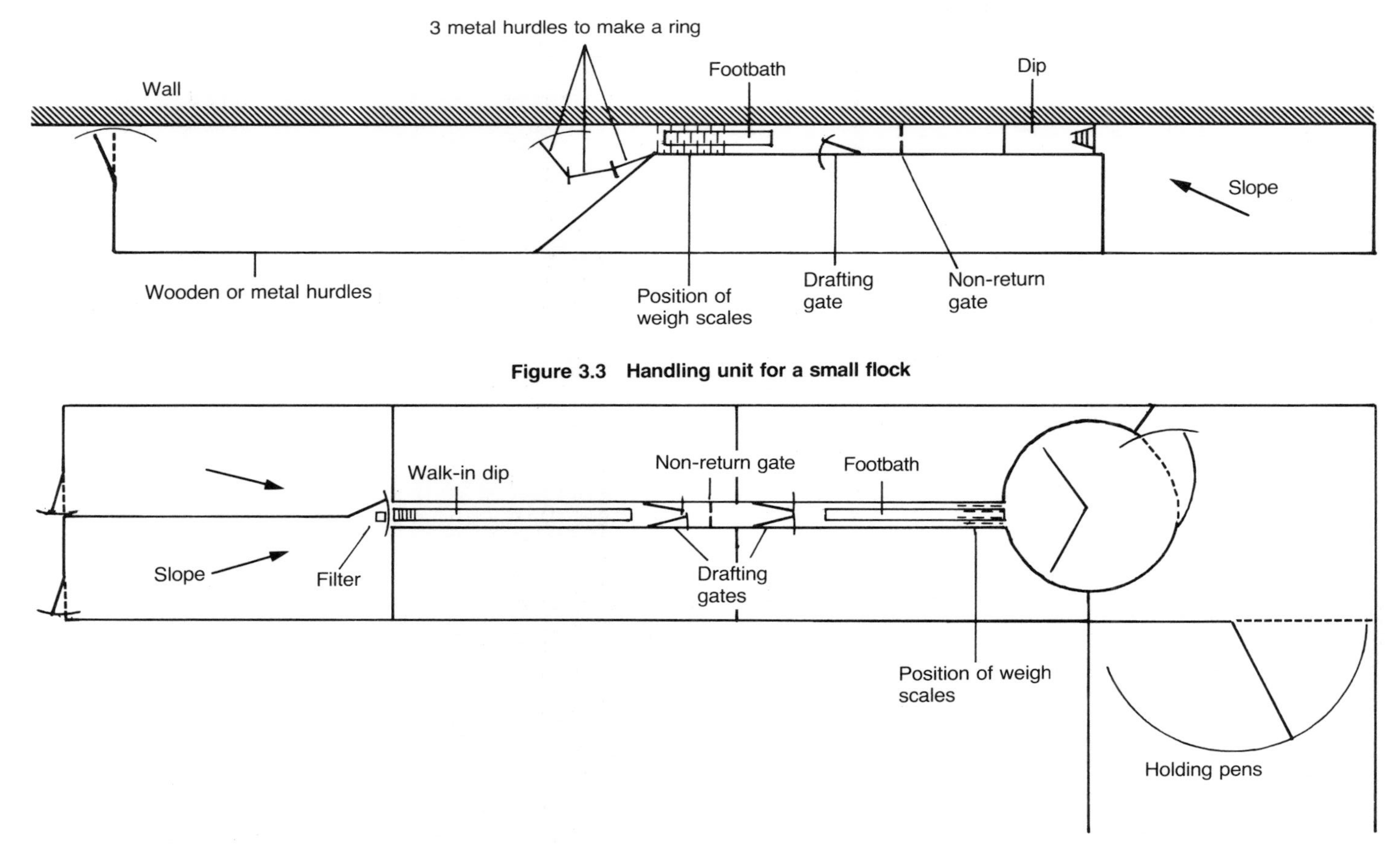

Figure 3.3 Handling unit for a small flock

Figure 3.4 Handling unit at Rosemaund EHF

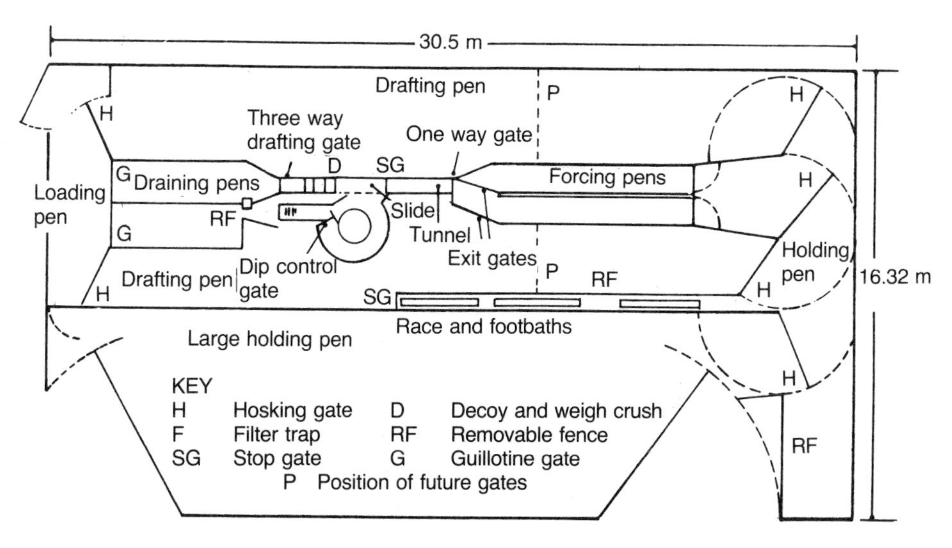

Source: *Farm Buildings Digest 12*, 3 Autumn 1977

Figure 3.5 Plan of the NAC sheep handling and dipping unit

Useful additions are a weigh crate for ewes and lambs, a handling cradle and a loading ramp. A roof over the working areas is very well worth-while, and an adjacent store is needed for the safe-keeping of drenches, dips and the many other materials used by the shepherd.

Before finalising the design, MAFF's *Codes of Recommendations for the Welfare of Livestock* should be studied. Also the MAFF leaflet *The Safe Use of Poisonous Chemicals on the Farm.*

Three designs to suit flocks of varying size are shown in Figures 3.3–3.5

Chapter 4

Feeding the Pregnant Ewe

Theoretical Aspects of Nutrition

The feeding of pregnant ewes is by no means an exact science. The nutritive value of feeds is widely variable and feed intakes are difficult to estimate, particularly where the flock is out-wintered. However, very considerable advances have been made in the last decade in our knowledge of the feed requirements of the in-lamb ewe. Many research workers have made contributions here, and none more than John Robinson of the Rowett Research Institute. The authors have drawn heavily on his findings and those of his co-researchers, and gratefully acknowledge their kind assistance.

It must be emphasised that the nutrition of the prolific lowland ewe is important the whole year round, whether she is dry, pregnant or lactating. Her feed requirement varies from being relatively low, when dry and again in mid pregnancy, to peaks just before lambing and in early lactation, and the adequacy of the feed regime is best indicated by the body condition of the ewe. The body condition-scoring technique is described in Appendix C. The aim should be to feed the lowland ewe to achieve various body conditions throughout the year which are indicated by the live-weight changes shown in Figure 4.1(a).

Although a balanced diet must always be offered, with adequate levels of protein, minerals and vitamins, the provision of energy in the feed is of paramount importance. The necessary energy inputs for the year, in terms of metabolisable energy (ME), are shown in Figure 4.1(b). These energy inputs may, of course, be achieved by many alternative feed regimes, with palatability and cost being important considerations. The composition and nutritive value of selected feeds are given in Appendix A.

Correct feeding of the pregnant ewe is an important key to flock performance and flock profitability for four reasons.

- It minimises foetal wastage, thereby increasing the likelihood of twins.

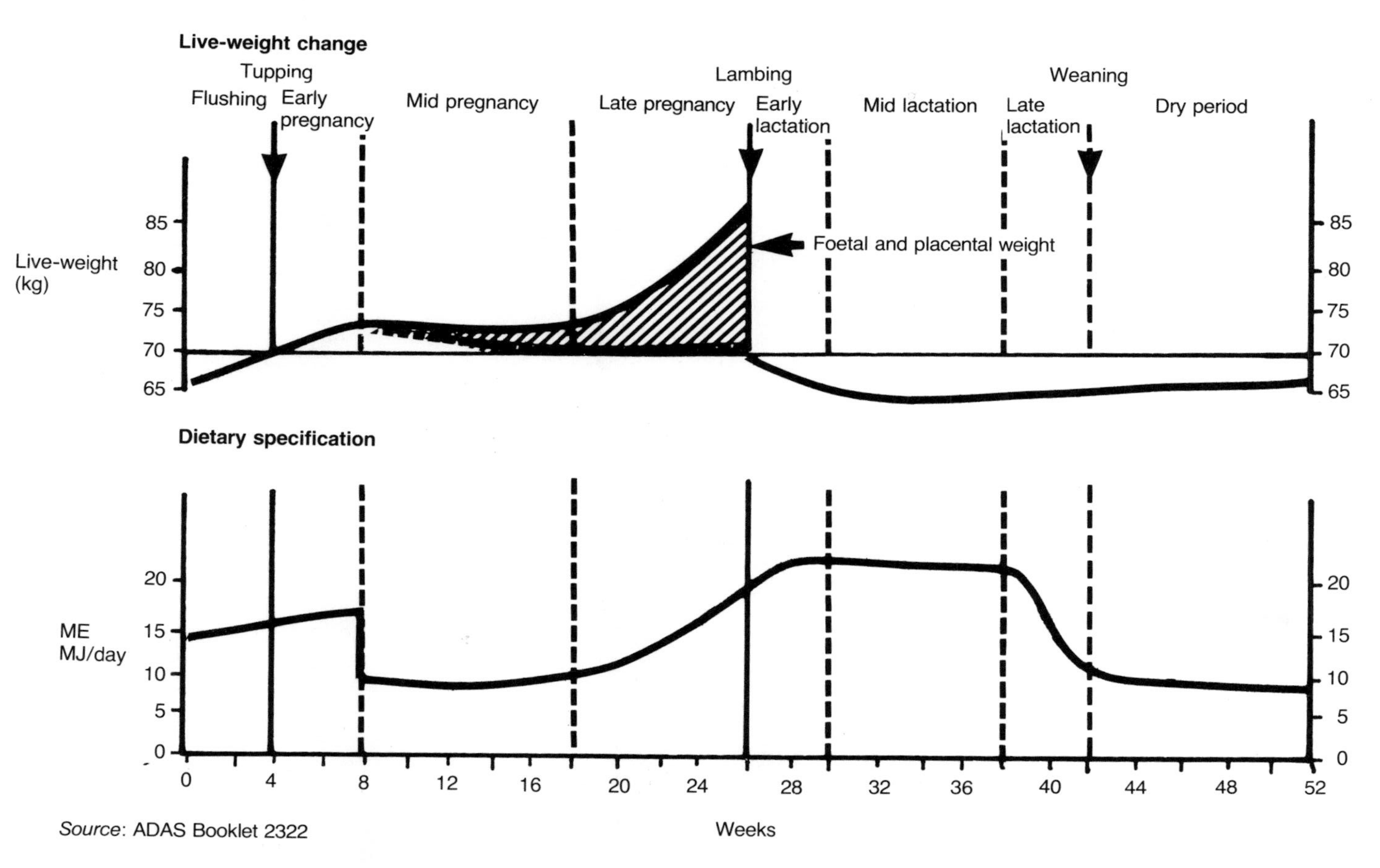

Source: ADAS Booklet 2322

Figure 4.1 Live-weight and nutritional standards for a 70 kg mature ewe producing twins

- It enhances lamb birth-weight, which means there are fewer small non-viable lambs.
- It safeguards ewe health, in particular reducing the incidence of twin lamb disease.
- It promotes maximum milk production.

The aim of the feeding programme must be to obtain the birth of a large number of lambs, these being vigorous and of a satisfactory birth-weight, at the least possible cost.

In the following pages we consider the nutritional requirements of the ewe in early, mid and late pregnancy (these are translated to practical feeding advice in Chapter 10) and review the advantages and limitations of the commonly used winter feeds.

Feeding in early pregnancy (first month)

The main objective is to prevent embryo loss. After fertilisation, the egg floats in the uterine fluid until it becomes attached to the uterine wall ('implanted') during the third week after conception. It is known that 15–30 per cent of the eggs shed at ovulation fail to develop into a lamb, and although some of these eggs may not have been fertilised, many more fail to implant. Much of this embryo loss results from stress caused by under- or over-nutrition and by abrupt changes in the level or nature of the feeding. The reaction of the ewe to any form of stress at this stage of pregnancy can be failure to implant a foetus.

The target body condition score of the lowland ewe at mating is 3.5, and it is now accepted that this level of fitness should be maintained over the first month of pregnancy. Therefore, the feed level at tupping time should be maintained unless poor body condition (below 3.5 body score) indicates the need for a little supplementary feed.

Feeding in mid pregnancy (second and third months)

By the end of the third month the growth of the foetus is small (approximately 15 per cent of the weight of the new-born lamb), and in a twin-bearing ewe total foetal weight is around 1.5 kg. This places little extra demand on the ewe for nutrients. In contrast, by day 90 the placenta has completed its growth process, and together with the thickened uterine wall and the uterine fluids it will weigh from 3 to 4 kg. The placenta, therefore, is much more vulnerable than the foetus to under-nutrition in mid pregnancy, as can be seen from Figure 4.2.

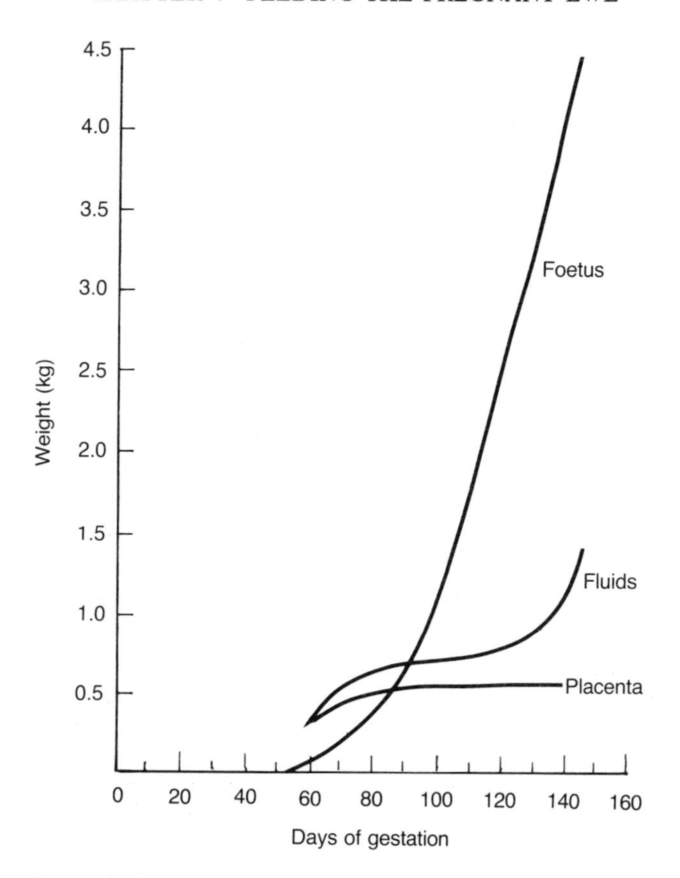

Source: J J Robinson 1982

Figure 4.2 The growth of each of twin foetuses and the associated placenta and fluids for a 70 kg ewe at mating

The implications for the flockmaster of the above are that ewes in fair condition (body score 3.5) should be maintained in live-weight in this period although a loss of up to 0.5 units in body condition is permissible. It is emphasised that there must be no period of severe under-nutrition since this will adversely affect placental development and, hence, the supply of nutrients to the foetuses.

Thin ewes must not be allowed to lose condition in mid pregnancy and should be separated off on better keep. Over-fat ewes should have been slimmed down between weaning and tupping, but any at a condition score of 4 or above should also be segregated and gradually slimmed down a little. Otherwise, such ewes are more prone to twin lamb disease in late pregnancy.

Feeding in late pregnancy (fourth and fifth months)

The feeding level at this time is critically important because it is a period of high foetal growth (70 per cent of this occurs in the last 6 weeks before lambing) and preparation for lactation. Feeding affects lamb birth-weight, lamb viability and the ewe's milk yield. Inadequate nutrition can result in horrendous lamb losses particularly from birth to 48 hours old.

It has been shown at the Rowett Research Institute that a modest energy deficit in late pregnancy is acceptable. However, this should not result in a loss of over 0.5 unit of body condition, and must not be accompanied by a deficit in protein intake. The flockmaster can exploit the inter-relationship between energy and protein by supplying some relatively undegradable dietary protein which may be in the form of fish meal or soya bean meal. This is beneficial when the diet fails to supply enough energy to meet the demands for foetal growth. The high-quality protein makes up the deficit in the last 3–6 weeks before lambing by enabling body fat to be efficiently mobilised.

Many writers have reported a considerable fall in feed dry matter (DM) intake in late pregnancy, from the normal level of around 2.5 per cent of the body weight of the ewe to about 1.7 per cent. At Rosemaund EHF, when high-quality grass silage (65 D+) has been offered, this fall-off in feed intake has not been observed. High-quality forages have the two advantages of encouraging improved intake and having a higher energy content, and may even satisfy the ewe's nutritional requirements right up to lambing. This has occurred at Rosemaund, and 'all-silage feeding' of the pregnant ewe is discussed later in this chapter.

With most hays and silages it is necessary to feed a compound ration to the ewes in the last 4–8 weeks before lambing. The objective is to increase the energy concentration of the ration when the ewe's requirement is rising rapidly and rumen volume may be reducing because of the rapid foetal growth. Most flockmasters increase the level of cereal-based concentrate weekly in line with requirement, but in recent years 'flat rate' concentrate feeding has been investigated and these alternative compound feeding strategies are compared later in this chapter.

Out-wintered ewes may well require considerable extra feeding in the last few weeks before lambing, or their performance may

Table 4.1 Metabolisable energy allowances for pregnant ewes kept outdoors in megajoules per day

Live weight (kg)		*Empty*	*Weeks before lambing* 8	6	4	2	0
60	Single	7.8	8.0	8.8	9.8	10.8	11.9
	Twins		8.1	9.4	10.9	12.7	14.7
70	Single	8.8	8.9	9.9	10.9	12.1	13.4
	Twins		9.2	10.6	12.3	14.2	16.5
80	Single	9.8	9.9	10.9	12.1	13.4	14.8
	Twins		10.2	11.8	13.7	15.8	18.3

NB: For ewes which are housed, the ME allowance may be reduced by 1.2 MJ/day.
Source: ADAS Reference Book 433, *Energy Allowances and Feeding Systems for Ruminants.*

be disappointing. With housed ewes the feeding level is under close control and should more exactly satisfy the ewe's needs. Moreover, the housed ewe does not need to walk more than a short distance to seek her food, and she is in a drier, draught-free environment, and therefore her feed requirement is smaller. Her energy requirement may be from 6 to 9 per cent less than that of the out-wintered ewe: the precise figure is still in dispute (Table 4.1).

Checking feed intakes

This is desirable for in- and out-wintered flocks, and should be a regular feature of the management of ewes. In housed flocks it is recommended that sample weighings of feed offered and refused are made on perhaps one pen in four at intervals of 2 weeks in order to check daily feed intakes. With the use of feed analyses, the energy intake can then be compared with theoretical requirements as shown in Figure 4.2. These check weighings should continue until 4 weeks before lambing, when final feed adjustments can be made before the ewes start lambing down.

Feeds and Their Quality

Hay feeding

In many flocks, hay is still the staple winter forage providing most of the nutrients until the last few weeks of pregnancy. It is liked for its adaptability, particularly in the baled form, when it can be conveniently transported, stored without deterioration and easily fed to both out-wintered and housed ewes. Unfortunately, it is also renowned for its variability. There are three main reasons for this.

- Hay is made from a wide variety of sward types.
- Grass for hay is cut at various stages of maturity.
- The British weather is not consistently helpful in providing wind and sun and no rain while the crop is curing in the field.

The best hay is made from grass cut at the early flowering stage and baled and carted in dry weather. In practice, most crops suffer weathering losses and mechanical losses (loss of leaf) which may total 20 per cent of the DM yield at cutting, and of course losses up to 100 per cent are not unknown! Although colour and smell give a guide to hay quality, it is strongly recommended that for a more accurate assessment, stocks of hay should be sampled and analysed. Appendix A gives the nutritive values of good-, medium- and poor-quality hays in terms of energy and protein content.

Hay should not be offered to ewes in overhead racks because of possible eye damage from falling hay seeds and because this feeding method leads to much wastage of hay in the bedding. Scandinavian hay racks covered by small-mesh wire and set just above floor level are preferable.

Table 4.2 Daily rations for 70 kg twin-bearing ewes

				Stage of production			
Hay quality	*'D' value*	*ME* (MJ/kg DM)	*Feeds and allowances* (kg/head/day)	*Maintenance*	*Tupping*	*Late pregnancy*	*Early lactation*
Good	61	9.3	Hay	1.1	2.1	1.3	1.9
			Compound feed	—	—	0.6	0.8
Moderate	57	8.5	Hay	1.2	1.9	1.1	1.5
			Compound feed	—	0.3	0.8	1.2
Poor	51	7.5	Hay	1.3	1.5	0.9	1.1
			Compound feed	—	0.6	1.0	1.5

NB: Compound feed assumed to contain 88 per cent DM and have an ME value of 12.5 MJ/kg DM and should contain 14–16 per cent crude protein (CP) depending on the results of the hay analysis.
Source: ADAS Leaflet 448, *Hay: quality and feeding*.

The considerations in feeding hay to pregnant ewes may be summed up as shown in Table 4.2 and as follows.

- Very good hay (ME over 10.0). An excellent feed requiring supplementation only in the last few weeks before lambing.
- Moderate hay (ME 8.5–10.0). Will satisfy maintenance needs. Should be rationed and supplemented at tupping time and throughout pregnancy.
- Poor hay (ME below 8.5). Has little place in the feeding of the prolific lowland flock. Can be fed (if not mouldy) but should be regarded as an alternative to cereal straw and supplemented accordingly.

Silage feeding

Sheep will not eat poor silage, but in the authors' experience they will eat palatable silage readily, and in fact will take to it even faster than cattle.

The quantity of silage made has increased dramatically in the last 20 years, and in 1980 it replaced hay as the premier grass conservation technique in the UK in terms of tonnage of DM. This is no surprise because the list of advantages of ensilage over haymaking is impressive.

- The feed value of silage is, on average, superior to that of hay and also more consistent.
- Ensilage allows more grassland intensification (e.g. higher nitrogen use) and results in earlier aftermath growth.
- It is less weather-dependent.
- The making and feeding of silage is more amenable to mechanisation. This is costly, but it does lead to savings of labour.

Silage quality for sheep

Silages to be fed to pregnant ewes should approximate to the following specifications.

- Dry matter content of 25 per cent or over is preferred. High DM silages result in better intakes and performance, fewer foot rot problems in housed ewes and a lower bedding requirement. However, a DM content of 20 per cent upwards is acceptable, and in a relatively wet climate, Liscombe EHF has successfully fed silages within the 20–25 per cent range over many years.
- 'D' value of at least 65 (ME at least 10.4). Vitally important to

satisfy the ewe's high energy requirement, particularly just before lambing.

- Acceptable fermentation. This is an indicator of high feed value, and also essential in order to achieve good palatability and high intake.
- Ash content of 10 per cent or less, indicating a low level of soil and manure contamination. Higher levels prejudice a lactic fermentation and reduce palatability.
- Chop length of approximately 20 mm. Experiments at Liscombe and Trawsgoed EHFs have shown that the intake of precision-chopped silage is higher than that of longer length material. Observations at Rosemaund EHF have indicated greater wastage of longer material in the bedding.

The techniques involved in the making of high-quality silage are now fully understood, and the essentials are as follows.

- Cut grass at or before ear emergence for high energy content. Swards containing legumes make silages which give better ewe performance than all-grass ones.
- Adopt a wilting technique to achieve a lactic fermentation and a high DM content. Use a silage additive if a successful wilt cannot be obtained.
- Fill the silo fast and exclude air by careful plastic sheet management.
- Keep a tight silage face in the clamp to avoid nutrient loss during the feeding period.

Clamp silage is generally a higher quality feed than big-bale silage, being more reliable and having a longer shelf life. The longer length of the forage in bales can depress intake and result in an enhanced need for supplementary feed. In experiments comparing similar silages at Trawsgoed EHF, big bales resulted in 10–20 per cent lower intake per ewe than clamp silages. However, the best practitioners do make excellent big-bale silages, a good example being Pwllpeiran EHF, which has also researched big-bale feeders.

Silage-based rations

With the best silages ('D' value in the region of 65) the supplement necessary may be mineralised cereals. At Rosemaund EHF, no protein supplements have been fed with silage of this quality for the last 13 years with no apparent ill-effects on the ewes. Mineralised mixed whole oats and barley are fed at 250 g/day in the penultimate week of pregnancy and at 500 g/day in the last week before lambing. Good but not excellent silages should be supplemented with a 13–14 per cent CP compound fed at the levels indicated in Table 4.3.

Table 4.3 Rations for pregnant ewes

	Weeks before lambing	*6* (kg/day)	*3* (kg/day)	*At lambing* (kg/day)
40 kg ewe				
	Silage	2.4	2.4	2.4
	Compound (S)	0.1	0.2	0.3
	Compound (T)	0.1	0.3	0.5
55 kg ewe				
	Silage	3.3	3.3	3.3
	Compound (S)	—	0.2	0.3
	Compound (T)	0.1	0.3	0.5
70 kg ewe				
	Silage	4.2	4.2	4.2
	Compound (S)	—	0.1	0.3
	Compound (T)	—	0.3	0.6

NB: S = single foetus, T = twin foetus.

The above rations assume silage of 25 per cent DM, 10.0 ME, 90 DCP and good fermentation.
Source: ADAS Leaflet P839, *Silage Feeding of In-lamb Ewes*.

Feeding techniques

Silage is rarely fed to out-wintered ewes because of its transport costs and the probability of feed wastage. It should always be introduced gradually and fed fresh. This means offering a 24 hour supply at each feeding and removing all refused material from the troughs daily. Good-quality material should be fed to appetite with no restriction on supply, since silage rationing predisposes sheep to the risk of twin lamb disease and also requires an increased trough frontage of 450 mm/ewe. Trough design is important (see Chapter 3) and, with ad lib feeding, the trough frontage allowance may be limited to 150 mm/ewe.

Some flockmasters consider that the feeding of silage to appetite increases the incidence of prolapse in ewes. At Rosemaund EHF, careful recording has shown the problem to be no worse in housed ewes fed silage ad lib than in out-wintered ewes fed hay.

Silage can be self-fed to ewes where the clamp is close to the sheep house. Advantages claimed include lower machinery costs and improved ewe health because of the beneficial effects on the ewe's feet from walking on concrete to and from the silage face. However, self-feeding limits clamp height to 1.3 m, thus increasing building costs; fleeces may become wet and soiled and savings in labour are small. In most situations, easy-feeding (trough feeding to appetite) is to be preferred to self-feeding as a technique which maximises silage intake. A wide choice of machinery now exists for the transport of silage from clamp to house, and ADAS mechanisation advisers are well able to advise on equipment. Whatever the choice, it is important to check intake on a weekly basis.

Silage-fed ewes require less water than those fed hay. At Rosemaund EHF, the water requirement just before lambing was found to be 1.8 l/ewe/day. Bedding needs of silage fed ewes have been 30–40 kg/ewe over a 10-week housed period.

Feeding ewes on silage only

After almost 20 years experience of feeding silage to housed ewes at Rosemaund EHF, it was decided in 1979 (Table 4.4) to discover whether or not ewes could obtain sufficient energy from good-quality silage to maintain them throughout pregnancy. Since it was deemed to be a little risky to feed 500 pregnant ewes on silage only, 30 ewes

Table 4.4 Quality of silage offered

	1979	*1980*	*1981*
DM (%)	27	27	28
Estimated ME (MJ/kg DM)	10.7	10.3	11.1
Crude protein (% in DM)	17.5	15.3	17.3

Source: Rosemaund EHF.

were housed in each of the next 3 years and fed solely on silage, with minerals available from housing in January until they lambed in mid March. Contrary to expectation, the DM intakes of the ewes did not decline as pregnancy advanced, as shown in Table 4.5.

In 1979 and 1981, when the quality of the silage was exceptionally good, the intakes of silage were also high. In 1980, the lower quality of the silage resulted in a reduced level of intake.

It can be seen from Figure 4.3, that in 1979 and 1981 the ME intakes were normally above the theoretical requirement throughout pregnancy, but that in 1980 energy intakes were lower and were below the requirement during the last 2 weeks of pregnancy.

The results in Tables 4.4–6 show that the ewes lambed down satisfactorily and that in all 3 years the birth-weights of the single and twin lambs were adequate. In 1980 – the year when the energy levels dropped below the theoretical requirement – birth-weights of the triplet lambs were very low and lamb losses were higher.

Table 4.5 Silage intakes (kg DM/day)

Weeks before lambing	*1979*	*1980*	*1981*
9	—	1.0	—
8	—	1.2	1.3
7	1.5	1.0	1.7
6	1.5	1.2	1.3
5	1.4	1.2	1.4
4	1.7	1.3	1.4
3	1.6	1.7	1.4
2	1.5	1.3	1.4
1	1.5	1.2	1.5

Source: Rosemaund EHF.

Research work in Ireland has led to the conclusion that even the best silages require supplementation when fed to ewes in late pregnancy. However, the Rosemaund EHF results showed that, providing the silage is of excellent quality, it can be regarded as a complete diet for the ewes and needs no extra energy supplementation. Such silages (see Table 4.4) are fairly rare, and with all silages below this quality

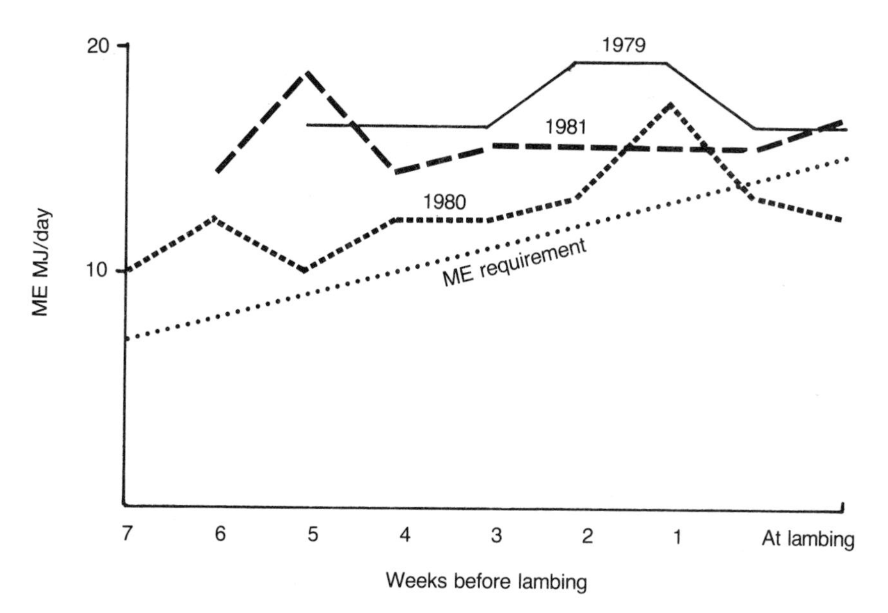

Source: Rosemaund EHF

Figure 4.3 Energy intakes of housed ewes on silage only

Table 4.6 Lambing results

	1979	*1980*	*1981*
Number of ewes	30.0	30.0	30.0
Number barren	—	—	1.0
Mean litter size	2.1	1.9	2.0
	Weight of lambs born alive (kg)		
Singles	6.0	5.2	5.5
Twins	4.3	4.3	4.5
Triplets	3.4	2.5	3.5

Source: Rosemaund EHF.

supplementary feeding is advisable because of the risk of twin lamb disease.

Silage and ewe health

There is no evidence that feeding silage to housed ewes poses any threat to their health, providing attention is paid to house ventilation, provision of dry bedding, adequate nutrition and foot care. Listeriosis is a disease which is associated with the feeding of silage. The reasons are not fully understood; poor fermentation and/or soil contamination are often blamed, and care must be taken not to offer material affected in such a way. Where listeriosis is a major problem, silage feeding to ewes should be avoided.

Straw-feeding

Out-wintered ewes will not willingly eat straw but for housed flocks, straw offers a genuine alternative to hay and silage as the main bulk feed. Straw-feeding is particularly attractive on arable farms (where this commodity has become an embarrassment since straw burning has come to be regarded as anti-social). Straw-feeding can mean the reduction or even abandonment of grass conservation, with consequent savings in mechanisation and labour. There is less pressure on grassland in the period from turn-out to July because there is no requirement for conservation, and this can lead to an increase in the stocking rate from the more traditional 10–15 ewes/ha to 25 ewes/ha or more. However, there are disadvantages, particularly the lack of grass aftermaths on which to graze weaned lambs and flush ewes before tupping. Straw-feeding is therefore most applicable to systems of production where lambs are sold by midsummer.

Feed value of straw

Straw is low in almost everything except fibre (see Appendix A) and is unlikely to provide even the maintenance requirements of ewes. Although perhaps more consistent in quality than hay, it depresses the energy and protein content of any diet of which it is a constituent. Normally, oat and spring barley straws are more digestible than winter barley, and wheat straws less so. The North of Scotland College of Agriculture has found variations in straw quality depending on variety, but more important may be the choice of clean, bright straws showing no mould growth. As the ewes prefer leafy material, it is sensible to choose the leafiest straw available regardless of variety.

The feed value of straw can be improved by mechanical or chemical means; chopping allows its incorporation with other feeds in 'complete diets'. When these are fed to appetite, the requirement of trough frontage per ewe is reduced from 450 to 230 mm. However, when straw is not mixed with other feeds, chopping is not advisable since it reduces the ewe's ability to select the leafier straws.

Chemical treatments include the use of sodium hydroxide or ammonia to break down cell walls and increase straw digestibility. In ADAS trials, successful treatment with sodium hydroxide has resulted in an average improvement of 15 units in 'D' value. Treatment with ammonia can give similar improvements, but results are more variable, and an average improvement of about 9 units in 'D' value is more likely. Chemical treatment can sometimes give disappointing results. In trials, poorer quality straws have shown a bigger response than better straws, and certainly in the case of the latter, money may be better spent on the purchase of high-energy, high-protein concentrates to be fed as supplements to untreated straws.

Straw-feeding technique

Ewes take time to adapt to straw-feeding. It should not, therefore, be introduced in late pregnancy, but rather fed for a period of at least 8 weeks, preferably being offered from the date of housing, with no preceding period of hay or silage feeding. The amount allowed should be not less than 1.5 kg/ewe/day (lowland ewes) and 1 kg/ewe/day (hill ewes) to allow for selective eating and for wastage. Fresh straw should be offered daily. Like hay, it is better fed in low-level troughs than in racks, although the feeding of big bales in suitable feeders has been successful.

Straw supplementation

Immediately straw is offered as the sole bulk feed it must be supplemented by a high-energy compound of at least 12.3 ME, with 16 per cent CP and an appropriate content of minerals and vitamins. Many purchased ewe nuts are not high enough in energy content and may contain milled cereals which introduce a danger of acidosis when the nuts are fed at the relatively high levels necessary. This is a situation where a home mix containing whole cereals and soya bean meal or fish meal is ideal, and the constituent percentages shown in Table 4.7 have been successful.

Table 4.7 Home-mixed supplement for straw-feeding

Constituent	*Amount* (%)
Whole cereals	77.5
Extracted soya bean meal	20.0
Sheep mineral/vitamin supplement	2.5

Where straw is the sole roughage fed to housed sheep, mineral deficiencies can occur. A high-phosphorus mineral supplement must be used, and a suitable specification is one containing 15 per cent calcium, 10 per cent phosphorus and 8 per cent magnesium; copper must not be included. Purchased compounds should contain at least 0.8 per cent calcium, 0.7 per cent phosphorus and no more than 0.35 per cent magnesium.

Supplementary compound should not be fed at over 500 g/ewe/feed, and (as with hay feeding) an adequate water supply is important, since ewes eating straw may drink up to 10 l water/day in late pregnancy. The suggested daily compound allowances to supplement straw are given in Table 4.8.

Straw-feeding at Rosemaund EHF

Early experience was with barley straw treated with liquid ammonia. A small rick of straw bales was built on top of a plastic groundsheet. A plastic oversheet was then positioned over the rick so that its edges

Table 4.8 Compound allowances for 70 kg straw-fed twin-bearing ewe (kg/day)

Weeks before lambing				
Over 8	*8–7*	*6–5*	*4–3*	*2–0*
0.45	0.65	0.80	1.00	1.05

Source: ADAS Leaflet P3016, *Feeding Straw to Housed Ewes.*

Table 4.9 Weight of compound fed during last 8 weeks of pregnancy

Weeks before lambing	*Silage*	*Barley or wheat straw*
	(kg/ewe/day)	
8	—	0.45
7	—	0.64
6	—	0.73
5	—	0.82
4	—	0.91
3	0.25	1.00
2	0.50	1.07
1	0.50	1.14

Source: Rosemaund EHF.

overlapped the undersheet. Soil was piled over the join, thus making a waterproof seal.

Liquid ammonia was then forced in under pressure using a lance which pierced the top sheet and some of the bales. The holes in the plastic made by the lance were sealed immediately the lance was withdrawn. A time lapse of roughly 6 weeks, depending on the temperature, was required before the straw was ready for feeding.

Initial scepticism concerning the feeding of treated straw was overcome when the ewes were seen to be tucking in eagerly. Performance (as predicted in theory) on treated barley straw was similar to that on moderate-quality hay.

With increased confidence in straw as a feed, the next step was to abandon straw treatment and evaluate untreated barley and wheat straws as the sole bulk feeds for pregnant ewes. The straws, suitably supplemented, were compared with grass silage for their suitability as feed for Mule ewes over the last 8 weeks before lambing (Table 4.9). There were 32 Mules in each group, and the experiment was conducted in 1986 and again in 1987.

The high-quality grass silage and the barley and wheat straws were

Table 4.10 Food and bedding costs during the 8 week housed period

	Silage plus compound (£/ewe)	*Barley straw plus compound* (£/ewe)	*Wheat straw plus compound* (£/ewe)
Silage @ £20/t	6.05	—	—
Barley straw @ £18/t	—	1.51	—
Wheat straw @ £14/t	—	—	1.18
Compound @ £127/t	1.11	6.10	6.10
Total food cost	7.16	7.61	7.28
Bedding @ £14/t	0.57	0.10	0.10
Total cost of feed and bedding	7.73	7.71	7.38

Source: Rosemaund EHF.

all offered to appetite. Ewes in all three groups received the same compound mix. This consisted of 775 kg of whole-grain barley, 200 kg of soya bean meal and 25 kg of proprietary mineral/vitamin supplements in a 1 t mix and had an ME of 12.3 MJ/kg DM and a CP of 16.2 per cent in the DM.

Ewes in all three groups lambed down at around 200 per cent, and lamb birth-weights and growth-rates were very similar in both years. The only significant effect of the treatments was that the ewes fed wheat straw lost more weight and condition in late pregnancy than ewes fed silage or barley straw. Food and bedding costs are illustrated in Table 4.10.

Two conclusions can be drawn from these experiments:

- Quality is more important than the type of straw. Good-quality wheat straw is preferable to weathered, mouldy, barley straw. However, good barley straw is better than good wheat straw.
- As far as cost or ewe performance is concerned it matters little whether the ewe is fed straw or silage. The recommendation is therefore that the flock is fed the food which the farm can supply best. On an all-grass farm, where straw is expensive to buy, silage is an excellent and economic food for the ewe. However, on arable or mixed farms where straw is available, its use is an excellent alternative.

Similar successful straw-feeding results have been reported by Trawsgoed EHF and from flocks monitored by ADAS in the south-east of England and on the Cotswolds. Straw-fed flocks recorded by the MLC had marginally fewer lambs and slightly higher replacement costs than on other feeds, but with improved stocking rates had gross margins per hectare similar to other flocks.

It appears that straw can replace hay or silage as the sole bulk feed for housed ewes with little effect on performance or feed cost per ewe. But there can be dramatic advantages in the stocking rate of the flock and in total farm income.

Hay, silage or straw?

The attributes and disadvantages of these three bulk feeds have been discussed and a choice must be made on the basis of these arguments. In the authors' experience, nothing is to be gained by feeding silage *and* hay or silage *and* straw. This is because the ewes will invariably eat good-quality silage and ignore the straw and all but the very best hay.

The choice of bulk feed has a big effect on compound requirements. This is illustrated in Figure 4.4, which shows the period before lambing during which ewes eating 1.5 kg DM/day should be fed compound. Where straw is fed, the DM intake is likely to be below this level. This figure confirms that the best silage provides for energy requirements through to lambing; also, that the traditional period of compound feeding commencing 6 weeks before lambing is correct for ewes fed average-quality hay. Straw should be supplemented from the time of housing the flock.

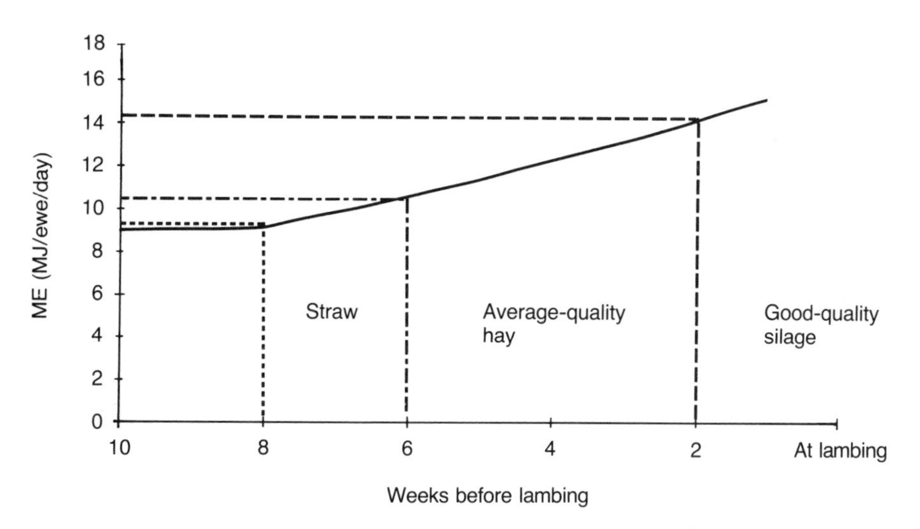

Figure 4.4 Using the quality of the forage to determine when compounds are required

Feeding forage crops

Catch crops such as fast growing turnips and forage rape have the ability to provide leafy keep in as little as 10 weeks after sowing and can be folded by ewes in early and mid pregnancy. However, hay should be available on a 'fall back' area together with low-level concentrate feeding, and on many of our heavier soils folding cannot be considered. The invention of electrified wire fencing reduced the labour requirement involved, but many flockmasters prefer to set stock sheep on roots, believing that this simpler management does not cause any additional feed wastage.

In late pregnancy, root crops should not normally constitute more than 50 per cent (66 per cent in the case of fodder beet) of the DM content of the ration. This is because their high water content limits energy intake, and high, not low, energy concentration in the diet is needed at this time. Comparison of the winter-hardy root crops suitable for feeding from the turn of the year has clearly demonstrated that in the lowlands, fodder beet outyields swedes and kale with ease. Modern varieties of fodder beet can be stored until May with little wastage to give a palatable high-energy feed; therefore, this crop has been investigated for its suitability, along with other feeds, for feeding to housed ewes right through to lambing.

Fodder beet for the pregnant ewe

Fodder beet is a high-energy root crop and has been evaluated at Rosemaund EHF for its suitability to form the whole or a part of the bulk ration of the pregnant ewe. In 1979, a group of 30 Mule and Welsh Half-bred ewes was offered chopped fodder beet to appetite plus 200 g soya bean meal/day with minerals and vitamins from the time of housing in January. After 6 weeks, the ewes were scouring and the bedding could not be kept dry even though twice the normal amount of straw was being used. It was thus concluded that all-beet feeding was a non-starter.

In 1980, and again in 1981, the following three rations were compared using pens of 30 ewes in each, from housing in January to lambing in March.

- Grass silage fed to appetite.
- Half silage and half fodder beet (on a DM basis) fed to appetite.
- Half hay and half fodder beet (on a DM basis) fed to appetite plus 200 g soya bean meal/ewe/day.

Performance on all diets was good, with a mean litter size of over

1.9 from each treatment group, and mean lamb birth-weight of twins from each group of 4.3–4.5 kg. The replacement of half the fodder beet by either silage or hay plus 200 g/ewe/day soya bean meal gave a completely satisfactory feeding regime for pregnant ewes.

In 1982 and 1983 the three rations were compared with the addition of a fourth treatment. This extra group of 30 ewes received half their ration of DM as fodder beet, and half as aqueous-ammonia-treated barley straw with the addition of 200 g soya bean meal/ewe/day. Ewe and lamb performance was satisfactory on all treatments, but blood analyses showed in both years that supplementary barley feeding was needed in the last 2 weeks before lambing for the ewes on the half and half fodder-beet and treated-straw treatment. This was because energy intakes were dangerously low, and there was a risk of twin lamb disease. In the 1982 trial, fodder beet was fed whole. However, it proved impracticable to sprinkle the soya bean meal over whole beet, and for this reason the beet was chopped before feeding in subsequent years.

Following the successful results from the feeding of fodder beet and treated straw, this ration was fed again in 1984 and compared with fodder beet and untreated straw (half and half on a DM basis) and with silage. Soya bean meal was fed with the two fodder-beet and straw rations as in previous years. Once again, lambing performance and birth-weights and growth of lambs were good on all these rations.

Fodder beet has several advantages as a feed for pregnant ewes.

- It is a reliable and high-yielding crop in the lowlands and can be stored until May with little deterioration.
- It is readily eaten by ewes.
- Ewes fed fodder beet normally have high intakes of energy, so supplementary cereal feeding requirements are low.
- It has a low fibre content and can be fed successfully with fibrous feeds such as hay and straw.

However, the following points should be borne in mind.

- Soil contamination of the roots should be kept to a minimum, and beet should not exceed 66 per cent of the diet DM.
- Root chopping is unnecessary if the ewes' teeth are in good condition. However, chopping facilitates the mixing of concentrate feeds with clean beet. If chopped beet is soil-contaminated, the supplementary soya bean meal or fish meal should be fed in a separate trough to avoid excessive soil intake and/or wastage of the protein feed in rejected soil cleaned out of the troughs.
- Beet is low in protein content. Supplementation with 200 g/ewe/day

of soya bean meal or with 140 g/ewe/day of white fish meal is advised where beet is fed with hay or straw. Ewes find soya more acceptable than fish meal when fed as a straight.

Supplementary feed

Supplementary feed can take the form of a purchased complete compound (often in nut form) or a home mix containing a purchased protein/mineral/vitamin supplement.

The importance of energy in ewe nutrition has been emphasised previously. Unfortunately, purchased compounds vary widely in ME content from 8.5 (medium hay value) to 13.6 (wheat value). There is now a reliable method for determining the ME from analysis. This will confirm the ME of a compound which has already been bought, but will not help in deciding which to buy. Declared analyses are of limited value in assessing quality; a list of ingredients and inclusion rates is far more helpful. *A word of caution*: never be tempted to feed dairy cake to sheep. Its copper content may be lethal.

There is much to be said in favour of simple home mixes based on whole cereal grains, a high-grade protein 'straight' and an added sheep mineral/vitamin supplement. Whole-grain is well digested by sheep and preferable health-wise to the milled cereals present in many purchased feeds. Additional advantages are that, knowing the constituents, we can calculate the energy content and, of course, home mixes are normally cheaper.

The best silages, with a CP in the DM of over 13 per cent, may be balanced with mineralised cereals. Lower quality silages, hay and straw should be supplemented with 14–16 per cent CP compound, examples of which are given in Table 4.11.

Mineral and vitamin supplements

Purchased compounds and high-protein pellets supply adequate amounts when correctly fed, but if neither of these is included in the ration the recommendation is to add a mineral and vitamin supplement specially formulated for sheep at the rate of 25 kg/t compound. This will supply the major minerals (calcium, phosphorus and magnesium), plus the trace elements (including cobalt and selenium) and vitamins A, D and E. However, there are situations where additional minerals are necessary, e.g. when feeding straw (see earlier in this chapter) and where there is a history of a trace element deficiency on the farm.

If a trace element deficiency is suspected, this can be confirmed by

Table 4.11 Compound mixes for pregnant ewes

	14% CP (*for feeding with good hay or silage of 13% CP or less*)		16% CP (*for feeding with straw or poor hay*)	
	(1) or (kg/t)	*(2)*	*(1)* or (kg/t)	*(2)*
Whole-grain cereal	825	850	775	800
Extracted soya bean meal	150	75	200	125
White fish meal	—	50	—	50
Sheep mineral and vitamin supplement	25	25	25	25

Source: Rosemaund EHF.

blood analysis and your veterinary surgeon can recommend appropriate therapy.

Feeding wheat to pregnant ewes

In the mid 1980s, the price of wheat fell almost to the barley price level, and in view of the superior feed value of wheat many flockmasters asked how much of this grain could be included in the ewe ration.

In response to this question, a 3-year trial was conducted at Rosemaund EHF in which all of the whole-grain barley in the compound was replaced by whole-grain wheat. The treatments were compared during the last 8 weeks before lambing with 30 Mule ewes in each group. All were allowed barley straw to appetite. The two rations were as shown in Table 4.12.

Wheat contains more energy than barley, and energy intakes of the two groups needed to be equalised in this trial. Therefore, the barley-fed ewes were allowed 0.45 kg/day compound at 8 weeks

Table 4.12 Barley- and wheat-based compounds

	Barley compound (kg/t)	*Wheat compound* (kg/t)
Whole-grain barley	775	—
Whole-grain wheat	—	775
Soya bean meal	200	200
Mineral and vitamin supplement	25	25

before lambing rising to 1.14 kg/day at lambing. The wheat-fed ewes received compound at 95 per cent of this level.

No significant differences were recorded between the groups in ewe live-weight change, mean litter size, or birth-weight or growth-rate of lambs. Moreover, there were no signs of ill-health or digestive upsets in any of the ewes.

The amount of wheat to be included in the ewe ration will depend on the relative prices of wheat and other cereal grains. However, the commonly given advice – that up to one-half of the barley in the ration may be replaced by wheat – seems a little conservative.

Feeding Methods

Floor feeding of compound

Is the provision of troughs necessary for compound feeding, or can a pelleted concentrate be fed in the bedding on the floor without disadvantage? This enquiry came from a farmer in Kent, who one morning found himself in the middle of a sheep pen with a sack of sheep nuts on his back, surrounded by a sea of sheep and unable to move to the troughs. He had no alternative but to throw the nuts on the straw floor in order to escape. Naturally, this method of feeding appealed to him but he wondered how much he was wasting by feeding on the floor.

In fact, no one could give him a definite answer, so an experiment was set up at Rosemaund EHF. Normally, the ewes are fed whole-grain barley and soya in a trough, and this was compared with the floor feeding of purchased pellets made to our specifications. The comparison was between the feeding of barley straw plus a home-mixed compound in troughs and the feeding of barley straw plus purchased pellets on the bedded floor. There were 30 Mule ewes in each group.

During the 3 years of the trial, no differences have emerged between feeding pellets on the floor or whole-grain in a trough in terms of ewe performance. The only difference observed is that after eating the food from the troughs, the majority of the ewes were lying down and ruminating after 15 minutes. In the pen in which the ewes were fed on the floor, the ewes rarely lay down and appeared to be looking for food all day.

Searches were conducted for pellets in the bedding 15 minutes after the pellets had been scattered on the floor, but none were found. This suggests that the ewes must have continued searching, more in hope than anything else; at least it gave them something to do.

The authors' opinion of floor feeding has changed during the period of the trial – from branding it as sloppy farming to regarding it as sound husbandry. Apart from making the actual feeding process easier, scattering the pellets all over the pen floor means that shy feeders are not bullied and pushed away as from a food trough. It also allows easier planning of sheep pens because it has always been difficult to design a pen with adequate trough frontage without placing extra troughs in the middle. In other words, the provision of adequate trough space places inhibitions on sheep house design.

A disadvantage of floor feeding is that a pelleted compound has to be purchased to reduce wastage. A home-produced whole-barley-based mix is obviously cheaper. However, the absence of troughs reduces the investment in equipment per ewe.

Flat-rate feeding in late pregnancy

One disadvantage of feeding straw as compared with either silage or hay is that more compound has to be fed. As we have seen in Table 4.10, this does not have a significant financial effect, but it can have a great effect on the ewe.

If a large amount of readily digestible carbohydrate in the form of rolled barley is fed to ewes, it can cause an increase in the acid content of the rumen, and if this develops into the condition known as acidosis, the ewe will go off her food and look and feel ill. If such ewes stop eating when their foetal demands are high, they go into negative energy balance, and pregnancy toxaemia (twin lamb disease) develops. Unless such ewes are promptly treated they may go blind and either abort their lambs or die.

In order to meet the theoretical energy requirements of a 70 kg ewe carrying twins, up to 1.1 kg/day of compound has to be given in addition to straw as lambing approaches. This total weight is usually fed in two equal daily feeds of 550 g of compound.

Providing no ewes exceed the compound levels intended with stepped feeding, no problems should arise, but if a greedy ewe eats more than her share, acidosis and twin lamb disease can occur.

To overcome this problem, flat-rate feeding during the last 8 weeks of pregnancy has been investigated. By this means the highest levels of compound feeding reached in late pregnancy under stepped feeding are never attained.

In the normal stepped rate feeding system, 47.3 kg compound/ewe has been fed with straw during the last 8 weeks of pregnancy. In the flat-rate trial this total amount was divided by 56 (8 weeks × 7 days) so that each day the ewe received 845 g of compound. In practice,

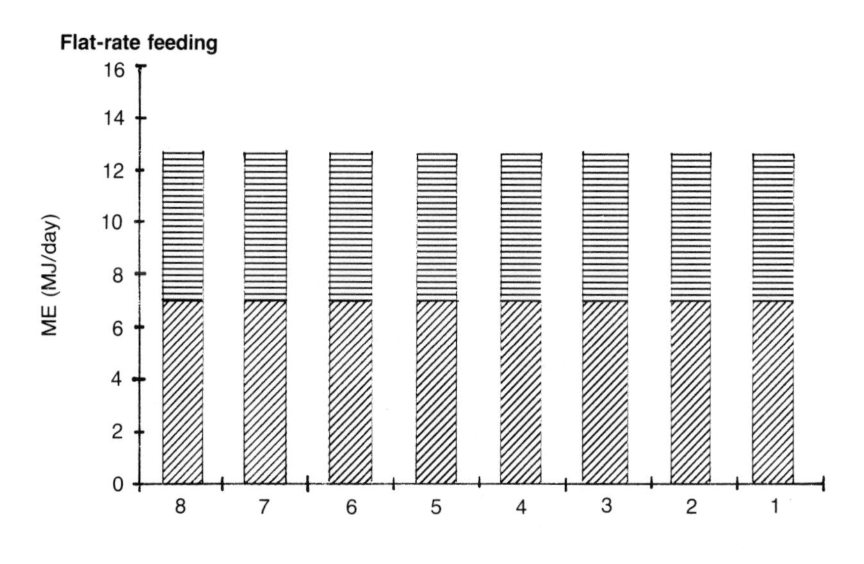

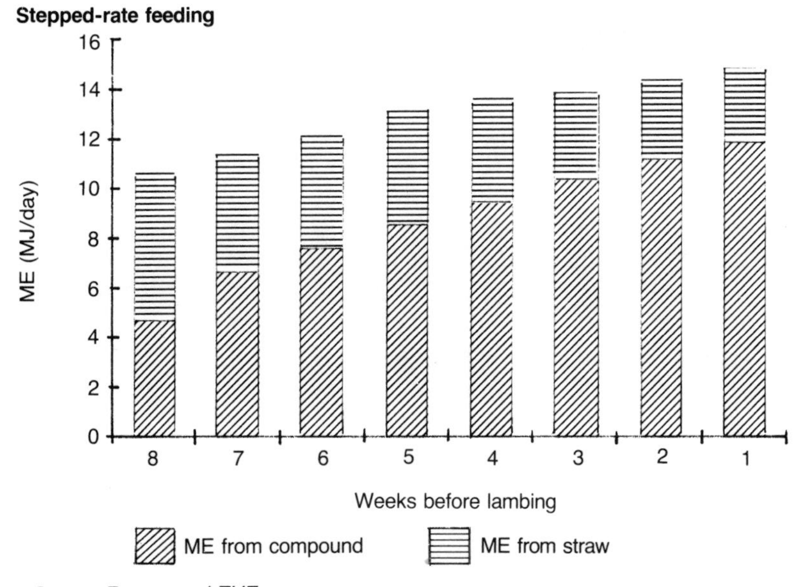

Source: Rosemaund EHF

Figure 4.5 Comparison of flat-rate and stepped-rate feeding of compound

the 845 g/day was gradually built up to during the first week. This investigation was carried out at Rosemaund EHF over 3 years. It involved 96 Mule ewes offered untreated barley straw to appetite, and compound either increased in steps or fed at the flat rate (Figure 4.5).

Ewes which received their compound in gradually increasing quantities were in energy balance at the end of pregnancy, but the ewes that were flat-rate fed ran into negative energy balance 2 weeks before lambing was due. The latter ewes should have developed twin lamb disease, but in the 3 years of this experiment no case of the disease has occurred, and the flat-rate fed ewes have performed as well as those fed in stepped amounts. Although ewes have tended to lose more weight after lambing, flat-rate feeding does appear to be a way of preventing acidosis, particularly in flocks fed straw as the only roughage.

Previous experiments on flat-rate feeding conducted at the East of Scotland College of Agriculture found that this method reduced metabolic disorders and the incidence of prolapse and was therefore to be recommended. Redesdale EHF was less enthusiastic. Blackface hill ewes when fed on the flat-rate system produced twin lambs which were as heavy at birth but grew more slowly than twin lambs from ewes whose compound ration had been 'stepped'.

Chapter 5

Grassland Management

High-level individual performance at grass can be combined with high output per hectare without high costs if the correct management decisions are made. Research workers and leading flockmasters have pinpointed the following key decisions.

- Accept the superiority of perennial ryegrass/white clover leys for grazing sheep.
- Optimise fertiliser applications and stocking rate.
- Control parasitic worms. Clean-grazing techniques and strategic dosing have made this possible.
- Keep grazing management simple and cheap by opting for set stocking or simplified rotational grazing.
- Realise that there is a case for the creep feeding of lambs.
- Get the weaning date right (and not too late!).

Two 'optional extras' should also be mentioned.

- Graze ewes with singles and ewes with twins separately (but there should be very few singles in lowland flocks!).
- Consider growing forage crops such as turnips. They integrate well with grass in the late summer.

Which is the Best Grass for the Ewe Flock?

Where permanent productive pastures exist on the farm (particularly those with a significant content of perennial ryegrass) they should be regarded as assets not to be lightly discarded. Such swards, compared with leys, are more resistant to poaching and drought, and at moderate fertiliser levels up to 200 kg N/ha they can equal many temporary swards in forage yield.

Short-term leys

One- and 2-year leys fit well into arable rotations. Italian ryegrass will not normally persist beyond 2 years but over this limited life-span it will outyield all other grasses. Red clover is not recommended because it has been known to reduce the fertility of grazing ewes. Short-term leys are valuable in reducing parasitic worm infestation. However, it should be remembered that Italian ryegrass does not react favourably to persistent grazing, to which it responds by dying out. It should therefore receive a cutting/grazing type of management. For those who prefer a ley of 3 or 4 years duration the choice lies between early flowering perennial ryegrasses and one of the hybrid (perennial ryegrass × Italian ryegrass) strains.

Long-term leys

There is a wide choice of seed mixtures for long-term leys. Some flockmasters prefer a 'Cockle Park' type containing several grasses, legumes and herbs. However, there is no evidence that such a sward will outyield a simple perennial ryegrass/white clover ley. Perennial ryegrass is favoured in many situations because it is second only to Italian ryegrass in yield, and because the late heading multi-tillering strains such as Melle (diploid) and Meltra (tetraploid) are easy to establish and will stand intensive sheep grazing over a period of at least 5–10 years. Information on recommended varieties of grass is contained in the NIAB Farmers' Leaflet No. 16.

White clover is an invaluable constituent of long leys, being high in energy, protein, minerals and vitamins. There is ample evidence, in particular from the Institute for Grassland and Animal Production (IGAP), Hurley, Berkshire, that with increasing clover contents of the forage there is a higher animal intake. This results in better performance by lactating ewes and by their lambs both before and after weaning. Scare stories that white clover may impair ewe fertility can be safely ignored, as may warnings that white clover causes bloat. It is, however, advisable to introduce sheep gradually to a clovery sward by feeding hay before turn-out and by limiting the time of grazing to 2 hours on the first day.

Clover is also valued because of its ability (by virtue of the bacteria in its root nodules) to fix atmospheric nitrogen. Its effectiveness in doing this is, however, variable, but most trial results indicate that a good stand of clover may be worth anything from 100 to 200 kg N/ha.

The establishment and management of clover in a mixed sward is all-important. It should be broadcast (not drilled) before mid August at 3–4 kg/ha at a depth of around 5 mm in a fine firm seedbed which should be ring-rolled. White clover varieties available are described in NIAB Farmers' Leaflet No. 4. In the authors' experience the large-leaved strains are to be preferred where a cutting/grazing management is imposed because they are less easily shaded out by ryegrass.

Perennial ryegrass/white clover leys should be managed to encourage the persistence of the clover at around 20 per cent of the forage by weight. This is achieved by limiting nitrogen applications to 100–150 kg/ha, and applying the bulk of this before May and the balance after early August, thus avoiding the clover growth period. It also means keeping the sward short. Remember that fertiliser nitrogen does not kill clover, but it does stimulate grass growth. If the grass is not frequently defoliated, its shading effect will rapidly reduce the clover stand. When planning to sow new leys the reader is recommended to study ADAS Leaflet P2041, *Grass Seed Mixtures*. This publication lists seed mixtures for leys for all durations and purposes.

Grass potential and seasonality

Yields of 20 t DM/ha/year have been obtained by cutting heavily fertilised Italian ryegrass swards. However, it is estimated that the average lowland grass yield is 5–7 t DM/ha/year.

The seasonal pattern of grass growth is important in that it dictates forage availability from month to month. The shape of the grass growth curve varies depending on rainfall, temperature and management, but normally rapid growth in the spring is followed by slower growth in midsummer and a second (smaller) peak of growth in late summer. Figure 5.1 shows two growth patterns obtained at experimental husbandry farms, both using 375 kg N/ha/year.

Fertiliser Policy

Soil samples should be taken from all fields every 3 or 4 years and a fertiliser policy based on these analyses. Lime should be applied as necessary to keep the soil pH above 6.0. With present high crop yields a target pH of 6.5 has been set on many mixed farms in order that soil pH will not be a yield-limiting factor where a variety of arable crops are grown in the rotation.

The soil index for phosphorus and for potassium should be 1 or 2, and a 'maintenance' dressing containing these elements should be applied to ensure that the indices are in this range. On many soils this may mean annual applications of each element applied at any time of the year when ground conditions are suitable except the spring. The spring application of potash has the effect of making magnesium less

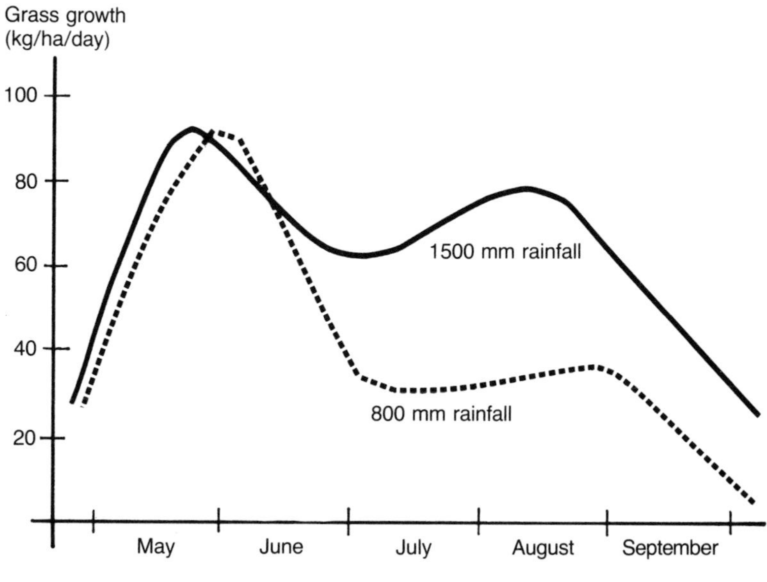

Source: ADAS Booklet 2052

Figure 5.1 Grass growth during the growing season (averages for 1970–76)

available, thus increasing the risk of staggers (hypomagnesaemia). Current recommendations are that phosphorus and potassium should be applied on a 'rotational' basis to those crops in the rotation most susceptible to their deficiency, and not necessarily every year. Advice should be sought on the frequency and level of these applications based on the results of soil analyses.

How much nitrogen?

There is ample evidence that a good grass ley will respond in yield in a linear (straight line) manner to applications of nitrogen fertiliser totalling up to 350 kg/ha over the growth season. It is also clear that such high-level fertiliser use has no adverse effect on the health of grazing sheep providing, of course, that suitable measures are taken to control internal parasites where sheep are densely stocked. Nevertheless, such a policy is appropriate only on pure Italian ryegrass leys.

The results of trials carried out at Rosemaund EHF in 1968–70 illustrate the effects of various levels of nitrogenous fertiliser applied to perennial ryegrass/white clover leys grazed by ewes and lambs at several stocking densities. The grassland received adequate maintenance dressings of phosphate and potash, and the nitrogenous fertiliser was applied in three dressings during the spring and summer. In each of the 3 years of the experiment nine flocks of twenty-eight Welsh Halfbred ewes together with their lambs by Suffolk tups were established. Each flock rotationally grazed around six paddocks, being moved whenever it was considered necessary. Paddocks surplus to grazing requirements were conserved to provide the silage fed to the ewes during the following winter. The grass-topping machine was used to remove unpalatable herbage and seed heads.

The nine treatments compared resulted from the combination of three levels of nitrogen fertiliser (0, 113 and 226 kg N/ha) with three stocking rates as shown in Tables 5.1 and 5.2.

Table 5.1 Weight of lamb sold and percentage finished off grass (3 years average)

Treatment	*Nitrogen rate* (kg/ha)	*Stocking rate* (ewes/ha)	*Lamb sold* (kg/ha)	*Lambs sold fit off grass* (%)
1	0	8.9	489	93
2	0	11.1	585*	81
3	0	13.3	702*	70
4	113	11.1	508	86
5	113	13.3	701	64
6	113	15.6	829**	55
7	226	13.3	724	91
8	226	15.6	822	74
9	226	17.8	894	58

* Insufficient silage made every year.
** Insufficient silage made in drought year.
Source: Rosemaund EHF.

Table 5.2 Effect of nitrogenous fertiliser on sward clover content

Nitrogen application (kg/ha)	*Percentage of herbage (ground cover) after 3 years* (white clover)	(sown grasses)	(unsown species)
0	35	50	15
113	9	69	22
226	1	79	20

NB: Before the commencement of the trial white clover constituted approximately 25 per cent of the sward.
The white clover loss due to application of fertiliser nitrogen occurred very rapidly, mainly in the first year of the experiment.
Source: Rosemaund EHF.

The lambs were sold when fit at around 36 kg weight. Lambs unsold in September were removed as stores.

Several important conclusions can be drawn from the results of the experiment. Firstly, the highest rate of nitrogen application at 226 kg/ha to a mixed grass/clover sward was unjustified in financial terms and was counter-productive. This was because the higher commitment of working capital involved to finance the additional fertiliser purchase and the extra ewes stocked on each hectare nullified the slightly higher weight of lamb output achieved compared with 113 kg/ha/year. Moreover, this level of nitrogen application virtually eliminated clover from the sward.

Secondly, the application of 113 kg N/ha/year supported a high stocking rate of over 15 ewes/ha throughout the year (with the exception of one drought year). Lamb output was high and some clover was retained in the sward.

Thirdly, where no nitrogen was applied, almost enough forage was grown to keep 11 ewes/ha through the year (there was a small silage deficit in the winter). A clovery sward was retained and most of the lambs were sold fit off grass.

Fourthly, the financial analysis of the results showed that the most attractive options were treatments two and six. The conclusions on nitrogen level and stocking rate on a mixed sward are as follows.

- On leys with a good clover content, a stocking rate per hectare of around 10 ewes and their lambs can be supported throughout the year with no application of nitrogenous fertiliser. This relatively low stocking density is attractive in its low working capital requirement. It also results in the early sale of lambs before the seasonal price fall.
- A practical level of intensification is to a stocking rate of 15–17 ewes/ha achievable by the use of around 120 kg N/ha. This requires additional working capital, but has given consistently good results at Rosemaund EHF.

The pattern of nitrogen dressings

In devising a suitable pattern of applications, the two objectives are to ensure a fresh leafy growth of forage throughout the season and to avoid jeopardising the persistence of the clover. The best policy is to apply nitrogen little and often, with at least half the seasonal total going on early (before May) and late (from mid August). At Rosemaund EHF, a total of 120 kg/ha has been applied in four equal dressings each of 30 kg/ha as shown in Table 5.3.

Grass is very responsive to the spring dressing which increases

Table 5.3 Pattern of nitrogen applications at Rosemaund EHF

Month	*Purpose*
1 Early March	To stimulate early growth
2 Mid May 3 Mid June	To sustain mid-season growth when the lambs have an increasing appetite and the ley is in its growth 'trough'
4 Mid August	To extend the grazing season

the early surge of forage production with little damage to clover. The timing of this first dressing is important. The average date at which the soil temperature at 10 cm reaches 5.5 °C should be ascertained, since this is the temperature which triggers grass growth. The first nitrogen application should be made 2 weeks before this date. The temptation to fertilise earlier should be resisted, since it will result in losses of nitrates in the drainage water which is both costly and environmentally damaging.

The August nitrogen dressing is beneficial in stimulating forage growth for the flushing of ewes and for the finishing of store lambs. Trials conducted by ADAS have shown a good response to mid August applications, with decreasing responses through September, and little response to October dressings in most seasons.

Stocking Density

As the stocking rate is increased, total animal output rises. However, individual performance may decline, with lower lamb live-weight gains and a reduced proportion of lambs sold fit off the ewe. This is the result of a combination of intensified competition for forage and increased levels of worm infestation. The aim of grassland management must be to secure a high stocking density without suffering a severe decline in individual performance and there is evidence that many flockmasters are achieving this aim.

The MLC flock recording schemes have shown that output and gross margin per ewe have been similar from flocks intensively and extensively stocked. If intensification does not greatly affect individual ewe performance this, of course, makes the case for a high density of stocking on productive grassland. It is noticeable that there have been many more examples of successful intensification, into the range of 15–20 ewes/ha, since the widespread adoption of ensilage and the consequent availability of earlier aftermath grazing than was possible after haymaking.

The effects of increasing stocking rate on the gross margin per hectare are illustrated in Table 5.4 taken from the recording of the flocks on 40 farms in 1985.

Although high stocking densities can improve both the physical and financial performance of the flock it is true that they can create problems such as those shown below.

- Individual performance can suffer.
- There may be poorer self-sufficiency for winter feed.
- Disease risks are greater.
- More working capital is tied up in the cost of stock per hectare.
- Inputs of fertiliser and supplementary feed will be higher.

Many flockmasters are overcoming the first three of these problems by good stockmanship and grassland management. There is ample evidence that sensible increases in stocking rate will improve output and more than cover the increased inputs involved and therefore improve the financial performance of the flock.

Table 5.4 Effect of summer stocking rate on gross margin per hectare

	Upper 25%	*Middle* 50%	*Lower* 25%	*All flocks*
Stocking rate/ha	14.7	11.8	9.1	11.9
Gross margin/ha (£)	617	570	372	532
Nitrogen/ha (kg)	147	96	74	103

Source: The West of Scotland Agricultural College.

In the lowlands, a good clovery sward of 1 ha which receives 120 kg N/ha over the growing season should support 15 ewes and their lambs throughout the year. However, on average grassland, 180 kg N/ha would be needed to support this stocking density.

Grazing objectives

The aim is to combine high stocking rates and high lamb growth rates. This can only be done if the patterns of, firstly, forage availability and, secondly, the animals' forage requirements, are continually adjusted so that the two are matched throughout the grazing season. *Forage availability* will be affected by the type of grass, fertiliser applications, grazing techniques and conservation. *Sheep forage requirements* depend upon the wintering system, supplementary feeding policy, together with dates of lambing, weaning and sale.

Both over-grazing and under-grazing depress performance. Over-grazing limits forage intake and therefore reduces the rate of live-weight gain. Under-grazing is perhaps even worse. It indicates a sub-optimal stocking rate which allows the grass to go to head. This reduces the 'D' value of the forage which in turn leads to poor performance unless the level of supplementary feeding is stepped up.

By avoiding under- and over-grazing, we can achieve the aim of grazing management. This is to maintain, throughout the season, the type of sward the sheep prefer – short, dense and clovery. But how short is short?

Grass grazing height for sheep

There is a close relationship between sward height and the performance of grazing sheep. This has been demonstrated in numerous trials, including those carried out at the Institute for Grassland and Animal Production (IGAP) at Hurley, the Macaulay Land Use Research Institute (MLURI) in Edinburgh, and Liscombe EHF in Somerset. The measurement of sward height gives the flockmaster a pointer to stocking density decisions and puts a little more precision into the notoriously imprecise technique of grazing.

Grass height can be measured without undue cost using a simple ruler at forty random points in the paddock or field. Alternatively, use the purpose-made HFRO or ADAS sward sticks, but remember that accuracy still depends on the taking of an adequate number of measurements.

In the mid 1980s, both IGAP and Liscombe EHF studied the effects of sward height on lamb live-weight gain. Both grazed ewes with twin lambs and the results in Table 5.5 were recorded from turn out to approximately 12 weeks later. The IGAP compared four sward heights, and Liscombe EHF compared two.

At Liscombe it was concluded that short, leafy, weed-free, highly tillered swards were developed by grazing at 3–4 cm in the first half of the season. The higher stocking rates on these relatively short swards meant better grassland utilisation. At IGAP, it was noted that the lower herbage intake at the 3 cm height restricted lamb growth, and the ewes lost considerably in body weight.

The evidence is that the sward height should be varied through the season to prevent the grass from heading and to satisfy the flock requirements shown in Table 5.6.

Plate 5.1 Use of sward stick to estimate grass yield

Table 5.5 Effects of sward height on live-weights of grazing ewes and twin lambs

		Sward height			
		(3 cm)	(6 cm)	(9 cm)	(12 cm)
Ewe weight change (g/day)	IGAP	−188	−54	−42	−67
	Liscombe EHF	+147	+242	—	—
Lamb growth rate (g/day)	IGAP	210	275	260	265
	Liscombe EHF	230	250	—	—

Sources: IGAP and Liscombe EHF

Table 5.6 Recommended sward heights

Season	*Sward height* (cm)	*Justification*
Spring	4	Allows ewes to milk well although stocking density high
Summer	6	Good growth of high-quality forage with no seed heads
Autumn	7–8	Less danger of grasses heading and more forage needed for flushing
Late autumn	4	Grass must be grazed down to prevent winter-kill

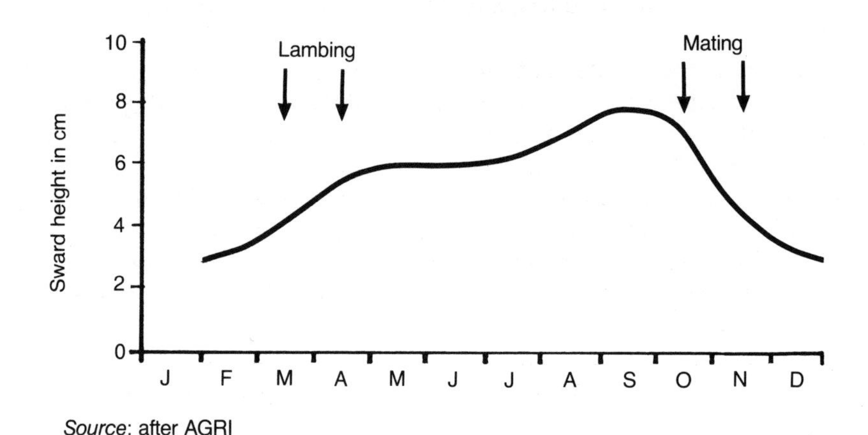

Source: after AGRI

Figure 5.2 Ideal profile of sward height for lowland sheep production

Grazing Management

There are several alternative methods of grazing management, and their strengths and weaknesses are set out in the following subsections.

Set stocking

The sheep have the run of the whole field throughout the season. This traditional method of grazing has four important advantages.

- It is simple. This commends it to the majority of flockmasters who dislike unnecessary complications.

- It is cheap. Only boundary fencing is needed, and just a few watering points.
- The sheep are allowed to graze the whole field with little disturbance.
- If well managed, it can give the best possible lamb live-weight gains.

Set stocking has two proven disadvantages and one which is unproven.

- Because it is difficult to assess grass growth under set stocking it is not easy to adjust the grazing pressure in order to avoid under- or over-grazing.
- Efficiency of grass utilisation may be lower than with rotational grazing. This can lead to a shortage of conserved forage for winter feed.
- It is often stated that set stocking enhances the risks from high levels of infective worm larvae on the pasture. In the authors' experience this is an unjustified criticism, since the alternative rotational grazing systems have little or no beneficial effect in reducing worm burdens. As explained later in this chapter, we must rely on clean-grazing techniques and strategic drenching to control parasitic worm infestation. A change from set stocking to paddock grazing on parasite-infested grass will do little to combat the worm problem.

If the flock is set-stocked it is very important to establish the optimum stocking rate through the season to keep pace with grass growth; providing this can be achieved, the advantages noted above should outweigh the disadvantage of rather poorer grass utilisation. So the advice on set stocking must be: 'Do it well and stick with it.'

The 'follow N' system

This system was developed by the sheep unit at the National Agricultural Centre, Stoneleigh, Warwickshire, and is a simple but effective development which has been taken up by many flockmasters. It is a variation of the set stocking system, and as originally devised, involves dividing the sheep pastures into four strips using marker posts erected in the boundary fences. One of the four strips is fertilised with nitrogen each week in rotation while the flock has free access over the whole field until the lambs are weaned. The dressings are adjusted to give the desired seasonal total.

The 'follow N' system is based on two principles.

- There is no health hazard involved in applying nitrogenous fertiliser to fields occupied by grazing sheep.
- The forage becomes temporarily less palatable when fertilised. This means that the flock grazes mainly on the strip fertilised 7–10 days previously, moving on to graze the other strips in turn at weekly intervals.

In recent years, the NAC has modified the system by reducing the strips from four to three, and a number of flockmasters are now using two strips only, alternately fertilised every 2 weeks.

The achievement of 'follow N' has been to secure some of the advantages of rotational grazing (better grassland management and utilisation) without the expense of fencing fields into paddocks.

Rotational grazing

In rotational grazing, the fields are fenced into a number of sheep-proof paddocks (usually between three and eight) around which the flock grazes in rotation. From the plant physiology point of view it is good management, since the sward is subjected alternately to a rapid defoliation and a period of around 4 weeks of complete rest.

Advantages

- Paddock grazing allows more accurate assessment of forage availability than does set stocking. There is therefore less likelihood of under- or over-grazing and the opportunity for better grassland utilisation.
- A higher year-round stocking rate may be achieved because of its built-in flexibility. One or more paddocks can at any time be closed up and conservation cuts taken, which means an improved supply of winter feed.
- It allows the forward creep grazing of lambs.
- Better control of internal parasites is claimed, but the claim is dubious (see above).

Disadvantages

- The bulk of forage on offer is variable, with a generous supply when the flock is turned into a fresh paddock, and very little on its last day of grazing. This leads to restlessness when ewes are impatiently awaiting a move.
- The costs of fencing, gates and water points are considerably more than for set-stocking. This extra cost may prohibit rotational

grazing altogether on short-term leys, because of the necessity to move all fences and water points every year or two.

- Additional labour is needed for stock movements and for fertilising small paddocks.
- Paddock grazing is a relatively sophisticated system, requiring frequent management decisions.

Set stocking vs. rotational grazing

Many trials have compared these grazing systems, and the Rosemaund EHF trials carried out in 1977–9 produced typical results.

At Rosemaund, perennial ryegrass/white clover leys were stocked by Welsh Halfbred ewes put to the Suffolk tup. The grassland received adequate phosphate and potash and nitrogen at the rate of 150 kg/ha overall. Half the total area of 6 ha was set-stocked and half split into six paddocks and rotationally grazed.

From turn out, two-thirds of each area was grazed, and one-third laid up for a silage conservation cut. The total area was grazed from the time aftermaths became available in June to the end of the season except where occasional cuts were taken of grass surplus to grazing requirements on the rotationally grazed area. The stocking rate per hectare was 22.2 ewes and their lambs in the spring and 14.8 from June onwards.

Lambs were sold when fit at approximately 38 kg live-weight. They were weaned at the beginning of August. Weaned lambs were given the best of the grazing, and those remaining in early September were removed as stores.

With a 9 per cent higher total lamb live-weight gain at grass

Table 5.7 Set stocking vs. paddock grazing

		Set stocking	*Paddock grazing*
Total weight of lambs at start	(kg/ha)	325	334
Total weight of lambs at finish	(kg/ha)	875	838
Total lamb gain at grass	(kg/ha)	550	504
Lambs finished off grass	(%)	71	64
Finished lamb output	(£/ha)	397	356
Store lamb output	(£/ha)	119	143
Total output	(£/ha)	516	499
Silage-first cut	(t DM/ha)	0.86	0.71
Silage-second cut	(t DM/ha)	—	0.92
Total silage	(t DM/ha)	0.86	1.63

Source: Rosemaund EHF.

Table 5.8 Set stocking vs. paddock grazing – financial balance

Set stocking disadvantages	(£/ha)	*Paddock grazing disadvantages*	(£/ha)
Cost of additional winter feed	46	Lower output	17
		Cost of fencing and water points	30
Total	46	Total	47

Source: Rosemaund EHF.

and a higher proportion of lambs finished off grass the set-stocking management achieved a total output which was £17/ha better than that from rotational grazing. However, the greater flexibility of rotational grazing allowed an extra conservation cut to be taken. With set stocking there was a deficit of forage for winter feed of 0.75 t of dry matter (DM)/ha. At £60/tDM cost, this debited the set stocking treatment by £46/ha.

With little or no financial advantage to rotational grazing showing up in most situations, it is not surprising that many flockmasters have opted for the simplicity of set stocking. Many others now operate a simplified rotational grazing system.

Simplified rotational grazing

Simplified rotational grazing means the grazing in rotation of whole fields; where, for instance, three fields of suitable size are available, two may be grazed from turn out and the third laid up for a conservation cut. Following this cut, the aftermath field is added to the grazing rotation, so reducing the stocking rate as the forage requirement of the growing lambs is increasing and grass growth declining. This grazing system combines the better grassland management inherent in rotational grazing with the low fencing costs which are a feature of set stocking.

Forward creep grazing

Forward creep grazing can be operated with any rotational grazing system. Developed at Cockle Park (Newcastle University) in the mid

1950s, it involves the installation of creep gates which allow the lambs (but not the ewes) to graze forwards into the next paddock in the grazing sequence. The objective is to give the lambs the first choice of fresh young grass, and to reduce the competition for forage which can occur between lambs and ewes at high stocking density.

Grazing studies at the IGAP, Hurley, showed that rotational grazing with forward creep gave a 13 per cent higher meat output and a 19 per cent financial advantage when compared with set stocking. Subsequently, in 1976–8, ADAS compared (on three farms in the south and west of England) a creep grazing system based on six paddocks with the normal system (set-stocking or two or three paddocks) on the farm. The conclusions were that forward creep grazing gave heavier lambs and the better lamb output per hectare, with carcass output per hectare from 4.5 to 16.9 per cent higher, but tended to produce more store lambs and fewer finished lambs.

Some flockmasters make a success of forward creep grazing, but the technique is not now widespread. The majority have concluded that the relatively marginal advantages in total lamb output which can be achieved are not worth the extra cost and complication.

Mixed stocking

For obvious reasons, the mixed stocking of dairy cows and sheep is usually inappropriate. However, grazing beef cattle and sheep together is traditional, and can be beneficial to both species as well as the grassland.

Why graze cattle and sheep together?

- Better pasture control. Cattle and sheep are complementary grazers and there is consequently less rejection of herbage.
- Sheep extend the grazing season in spring and autumn with less danger of poaching.
- There is ample evidence from Trawsgoed EHF and elsewhere that mixed stocking can result in higher total live-weight gain per hectare than grazing with either cattle or sheep alone.
- Mixed stocking may reduce (but not eliminate) stomach worm infestation. This is because most parasitic worms are specific to cattle or to sheep and the effect of grazing both species is to dilute worm infestation.

What are the snags?

- Competition between the cattle and sheep grazed together can delay lamb sales and reduce cattle live-weight gains.

- Fencing is more expensive, and the supplementary feeding of one species only is virtually impossible.

The conclusion is that the presence of sheep as subsidiary grazers to beef cattle is advantageous in 'cleaning-up' after the beasts and in grazing down the leys in wet autumns (when cattle would damage the sward) in order to avoid winter kill. However, many flockmasters achieve good grass utilisation and good live-weight gains without the unnecessary complication of mixed stocking.

The presence of beef cattle on the farm does, of course, facilitate the adoption of a clean-grazing system which is highly recommended. Here, the cattle and sheep graze pastures in alternate years, not together.

The separate grazing of twins and singles

Ewes suckling twin lambs are particularly susceptible to any shortage of feed in the first weeks after lambing because their energy requirement is around 25 per cent higher than that of ewes with single lambs. If feed availability is curtailed at this time, their milking ability is reduced; this has serious repercussions on the weight gain of their lambs because the milk intake of each twin is normally only about 70 per cent of that of singles. All this means that the nutritive requirements of ewes with twins and ewes with singles are different, and it is therefore sensible to accept this difference and to graze them as separate groups. The twin-suckling ewes should have the best grass and a lower stocking density, and supplementary feed should be considered for the first 4–6 weeks from turn-out. Calcined magnesite can also be added to the supplementary feed as ewes with twins are at risk from magnesium deficiency.

Twin lambs are known to eat creep feed earlier than singles and to consume more of it before weaning, so the creep feeding of twins and triplets is particularly rewarding; this is another reason for separating them from the singles, which have more milk and are less reliant on this extra food.

A final thought on this technique. In well-managed, prolific lowland flocks there should be only very few ewes turned out with a single lamb. Those giving birth to singles should be roughly matched in numbers by those producing triplets, and a good mothering-on technique should ensure that virtually every ewe has two lambs at foot at grass. So in the lowlands it is really an admission of defeat if this technique is an important element in flock management.

Supplementary Feeding of Ewes at Grass

Lamb growth during the first 6 weeks is dependent mainly on the ewe's milk supply, and so it is most important to establish good lactation. Milk yield is influenced by the size of the ewe (bigger ewes yield more), age (third to sixth lactations are the best) and number and size of lambs (ewes with twins may yield 40 per cent more than those with singles). It is also affected by the level of nutrition both in late pregnancy and in early lactation. Inadequate feeding in these periods will mean an early lactation peak (at 2–3 weeks instead of the ideal 3–5 weeks) and a lower total milk yield.

It is known that a ewe with twins has a 70 per cent higher energy requirement in the first month of lactation than she had before lambing. The question is: Will the best grass supply this need? Some flockmasters believe that it will, and that compound feeding at grass merely replaces grass in the diet and, moreover, unsettles the ewes. They spend their time looking for the next trough feed instead of foraging. However, most have found that even the best pastures may not supply the full needs of ewes with twins until late April in the south and mid May in the north.

The Rowett Research Institute found that an increased protein content in the compound fed to lactating ewes increased the milk yield. It is now clear that even when high-energy diets are fed there is a response in yield if the protein provided by the rumen microbes is supplemented by 'undegradable' sources of protein such as fish meal, extracted soya bean meal or linseed meal. The supplementary feed should therefore be high in energy and have a 16–18 per cent crude protein (CP) content to include relatively undegradable protein constituents.

Since the ewe's peak nutritional requirement is at 2–3 weeks after lambing, the level of supplementary feeding should be 0.5–1.0 kg/day at this time (depending on grass growth) with subsequent gradual reductions to nil by 6 weeks after lambing.

Creep-feeding Lambs at Grass

Where ewe flocks are grazed at high stocking densities, competition for the available forage builds up between the ewes and lambs. Since the ewes are the more successful competitors, this curtails the feed intake and progress of the lambs. However, any feed deficit may be made good by the provision of a suitable compound in a lamb creep-feeder inaccessible to the ewes.

The justification for creep-feeding is that it optimises lamb growth rate even where the stocking rate at grass is high. It is also claimed that it may reduce parasitic worm infection since creep-fed lambs are less dependent on grass, but there is no clear evidence in support of this claim.

Lambs will nibble their first creep feed at 2–3 weeks, and eat a significant quantity by 6 weeks. The compound should be palatable and high in metabolisable energy (ME), about 12.5, and can contain a coccidiostat. The practicabilities of the technique are described in Chapter 12.

Creep-feeding is only valid economically where it enables lambs to be marketed in the period before the lamb price drops at the end of June. It should not be regarded as a temporary stopgap measure, but should commence when lambs are 10 days old and should continue through to slaughter.

Alternative Forage Crops

Forage crops may be integrated with grassland in the grazing programme. They are particularly valuable in late summer when grass growth declines. At that time of year, lamb growth rates may fall and ewes decline in body condition, so jeopardising subsequent prolificacy. A catch crop of turnips or rape sown in April can provide invaluable keep 10 weeks later in late June/early July. This is ideal forage for weaned lambs and, moreover, should be worm-free; however, the lambs should be dosed with an anthelmintic on entry to the catch crop.

Lambs at Liscombe EHF have performed well on these forage crops when stocked at 35/ha. The removal of the lambs on to catch crops allows the ewes to build up body condition prior to tupping without competition from weaned lambs for the available grass.

Forage rye sown in the autumn will bring forward the grazing season in the spring by providing keep some 2–4 weeks earlier than grass leys and is particularly valuable for the early lambing flock. Winter feed can be provided by swedes, kale or cabbage. However, turnips and forage rape are the most attractive alternative forage crops, particularly as, following their grazing in the July–September period, the land can be sown to winter cereals. They do have the problem (as yet unsolved) that the grazing sheep cause considerable crop wastage. In trials at Liscombe, Redesdale and High Mowthorpe EHFs this wastage has varied widely with figures in excess of 50 per cent reported in wet conditions.

SHEEP HEALTH AT GRASS

Disease problems prevalent in sheep at grass fall into three categories: trace element deficiencies, metabolic disorders and internal parasites.

- *Trace element deficiencies*. These are numerous, but the most troublesome are:
  cobalt, causing unthriftiness in lambs and, in severe cases, pine;
  selenium, associated with vitamin E deficiency in causing white muscle disease.
- *Metabolic disorders*. Hypomagnesaemia (staggers) is the most important. Also hypocalcaemia (lambing sickness).
- *Internal parasites*. The four important ones are coccidia, liver fluke, lung worms and stomach and intestinal worms.

Routine preventative measures found to be successful against the above are described in Chapters 8–12, but these should be put into practice only after veterinary advice on the farm. However, the authors have thought it helpful to include a section on worm control in this chapter for two reasons. The first is that the level of worm infestation has an enormous influence on the growth rate of lambs at grass. The second is that in recent years it has been shown that a combination of strict grazing regimes and the strategic use of anthelmintics gives us better control of the worm menace.

Worm control in sheep

Much of the following has been taken from ADAS Booklet 2324, *Clean Grazing Systems for Sheep* and from reports published by the East of Scotland College of Agriculture.

The two types of worms which cause most problems are *Nematodirus battus* and the stomach worms, *Ostertagia. Nematodirus* worm eggs over-winter on the pasture and hatch into infective larvae when it is warmer. During a prolonged cold winter, these eggs may not hatch until April/May (just when lambs are starting to graze) resulting in severe *Nematodirus* outbreaks. This worm is passed on from one lamb crop to the next. *Ostertagia* cause the main worm problems in summer and autumn and these result in part from over-wintering infection from eggs dropped in the previous summer and autumn, and in part from eggs dropped by lactating ewes in the spring. As lambs become infected they also pass out worm eggs, the result being a peak in the pasture larval level in late June and July.

It is the study of the life cycles of the worms concerned which has enabled researchers to identify the risk periods when pastures should not be grazed. This has led to the definition of clean grazing as follows.

- *Clean grazing in the spring.*
  Newly established grass leys not infected by worm larvae.
  Pastures which have not carried sheep in the previous 12 months.
- *Clean grazing from July onwards.*
  Aftermaths not grazed by sheep since the previous autumn.
  Pastures free of sheep since the previous autumn.

It is also considered that pastures grazed only by dosed ewes in the previous autumn will be 'safe' for grazing by ewes and lambs, although not necessarily clean.

At weaning time each year, steps should be taken to make available clean grazing in the following season. *Next year's sheep pastures must not be grazed by this year's lambs.*

The simplest system is to graze beef cattle and sheep in alternate years. However, this does not give clean pasture for lambs at weaning, and one way of solving this is to adopt a 3-year rotation of cattle, sheep, and conservation/weaned lambs in that order, as illustrated in Figure 5.3.

The combination of clean grazing and drenching

Combined clean grazing and drenching is the policy which gives the most effective control of worm parasites.

- *In the spring:* dose all ewes (and any lambs over 4 weeks old) before turning them on to clean pasture. This will control pre-weaning infection, and dosing of lambs before weaning is then unnecessary.
- *At weaning:* dose all lambs and move them on to either clean grass or forage crops. This controls post-weaning infection.
- *In the autumn:* dose ewes if they graze next year's sheep pastures. Do not allow store lambs and hoggs on to next year's sheep grazing.

This proven policy of integrated worm control as developed at the East of Scotland College of Agriculture gives three big advantages to the intensive sheep grazier as follows.

- *Better lamb growth rates and earlier lamb sales.* In East of Scotland College trials lambs on this system have gained 300 g/day to weaning. Their gain was 38 per cent better than that of lambs on dirty pasture with regular anthelmintic drenching. Both groups were heavily stocked with 15 ewes and their lambs per ha. These results were confirmed in 1985 at the Macaulay Land Use Research

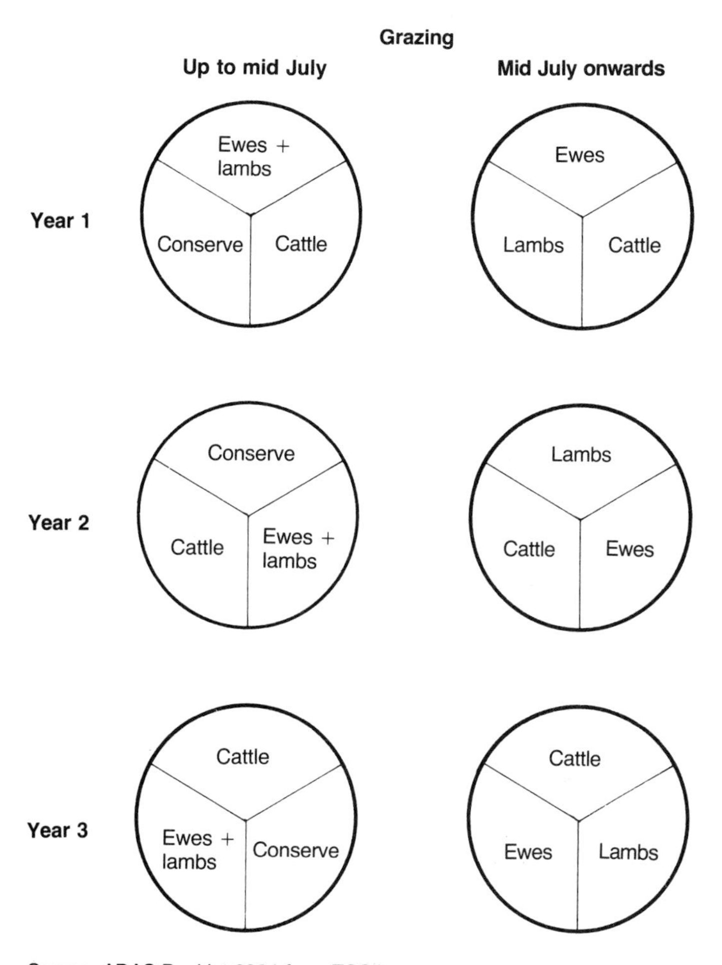

Source: ADAS Booklet 2324 from ESCA

Figure 5.3 The Rutter ring system

Institute's upland research station at Hartwood in Lanarkshire where lambs (drenched) on worm contaminated pasture had growth rates from 30 May to 16 July which were 30 per cent below those on pasture which was part of a clean-grazing system. This trial was reported by A. R. Fawcett and A. J. Macdonald in the *Vet Record (1988) 123.*

- *Higher stocking rate.* The keeping of more ewes on the same grassland area is made possible without a decline in performance, thus releasing land for other purposes. It also means that the sward is kept short which in turn means that it is of higher nutritive value because of its greater digestibility.

- *Less-frequent drenching.* The adoption of a clean-grazing regime cuts the number of anthelmintic drenches necessary to control worms. This reduces costs and labour requirement.

What if 'clean grazing' is difficult or impossible to find?

If there are no beef cattle on the farm, clean grazing for sheep is very difficult to find unless there is a high proportion of newly sown short-term leys. So a clean-grazing policy becomes impracticable. Even on beef/sheep farms some flockmasters have found difficulty in providing clean grazing in the early spring before the flush of grass growth. It is clear that a clean-grazing system will not fit in on all sheep units, and an alternative worm-control policy will then be needed. This places a greater reliance on regular dosing with anthelmintics. Rotational grazing gives no significant increase in worm control because of the long survival period of these parasites.

At Rosemaund EHF, good lamb performance has been achieved on 'dirty' grass by dosing ewes at housing followed by the regular dosing of lambs from May onwards. It appears that the flock can 'live with worms' if the frequency of drenching and choice of anthelmintics is good and if silage aftermaths become available in June; the Rosemaund routine is described in Chapter 12. It should be noted that it is recommended that flockmasters 'ring the changes' on the anthelmintics used. They are not all equally effective against the various worms and, furthermore, there have been reports of anthelmintic resistance building up. Over-use of the same drench might increase the likelihood of a resistance problem.

Whatever the worm-control policy adopted it is always worthwhile to follow the infestation forecasts of the Ministry's Central Veterinary Laboratory and the local Veterinary Investigation Centre, and to secure the detailed advice of your veterinary surgeon.

Chapter 6

Finishing Store Lambs

In the UK, the 'natural' time for lambs to be marketed is from August to October, and there is a glut during these months. However, demand for lamb exists the whole year round. This is recognised in the UK guide price regime, which has as one of its objectives the orderly marketing of lamb by encouraging sales outside the glut period. Traditionally, imported New Zealand lamb has supplied much of the UK market during the first half of each year, but that country has made efforts to diversify into other markets.

The UK level of self-sufficiency is rising, as is our ability to supply 'out-of-season' lamb. This may be done in three ways. Firstly, we can store carcasses by refrigeration, but the storage costs are high, making this option unpopular with the trade. Secondly, we can breed ewes outside the normal breeding season; the techniques are described in Chapter 7. Thirdly, we can put lambs through a 'store' period of low live-weight gain, in order to delay their marketing until a lamb shortage occurs and prices rise.

Because of the high costs of refrigeration and the difficulties experienced in maintaining prolificacy with out-of-season breeding, the finishing of store lambs is by far our most important means of levelling-out the lamb supply. Indeed, it has been estimated that around two-thirds of our lambs (including most lambs from hill and upland flocks) are subject to some form of store period.

The lowland flockmaster can regard the finishing of store lambs as an additional opportunist enterprise. He or she may have the remains of the lamb crop (the lambs which have failed to reach marketable condition by the autumn, and which it may be preferable to finish at home rather than risk the vagaries of the store-lamb market). He can also purchase store lambs which are available in large numbers from July onwards. This can involve a substantial commitment of working capital – but for a relatively short period – ranging from as little as 5 weeks to as much as 5 months. The type of lamb to be purchased and the appropriate feeds will depend on whether short- or long-keep systems are chosen.

Short-keep finishing

The objective here is to achieve a rapid finish in 6–10 weeks by ensuring a live-weight gain in excess of 1 kg/week. Down-cross ewes and wether lambs by Suffolk or continental breed rams are suitable; also first draw halfbred wether lambs out of ewes of the hill breeds by longwool sires such as the Border Leicester and Blueface Leicester. They are available at store-lamb sales held in August and September, and among the most popular crosses are the Scots and Welsh Halfbreds, Greyface, Mule and Masham.

Skill in buying is all-important for short-keep profit, since no significant rise in lamb price per kilogram can be anticipated in the August – November period. Cheap homegrown forages are the normal feeds, either aftermath grass, stubble turnips, forage rape or sugar-beet tops.

Long-keep finishing

Long-keep finishing is based on the theory that it pays to buy relatively cheap lambs in the autumn and hold them until the rising markets of February and March. Late-maturing lambs of the hill breeds and their crosses, e.g. the Scottish Blackface and North Country

Plate 6.1 Scottish Blackface store lambs eating grass silage

Cheviots, are good choices. Such lambs marketed by the Highlands and Islands Development Board have given highly satisfactory results at Rosemaund EHF. Also suitable are the large-framed lambs sired by longwool and down breeds such as the Bluefaced Leicester, Suffolk and Oxford. These may be held for up to 5 or 6 months for eventual sale in February or March at carcass weights of up to 26 kg.

The aim in long-keep finishing is to grow frame and maintain condition on low energy diets until late January or early February when supplementary compound feed is introduced for the final 4–6 weeks of finishing. Until Christmas, the lambs may graze cereal stubbles, surplus grass, stubble turnips, rape or sugar-beet tops, to be finished from the New Year onwards on hay or silage. Alternatively, whole-season root crops may be favoured because of their winter-hardiness. Lambs may be fed turnips followed later by sugar-beet tops, swedes or the leafy brassicas such as kale or cabbage.

Alternative Feeds for Lamb Finishing

Aftermath grass

Aftermath grass remains the most popular feed for the finishing of store lambs and is ideal for the purpose, particularly on dairy farms where the pasture is free from worm infestation. The best results are obtained on grass/clover aftermath swards given a dressing of nitrogenous fertiliser at the rate of 30–60 kg N/ha during August. Such grassland will keep 20–30 lambs/ha for a period of 6-10 weeks, giving from 1000 to 2000 lamb grazing days per hectare. Poorer quality grassland or swards unprepared for aftermath grazing commonly provide only about half the above number of grazing days.

It is always wise to reserve the best autumn grass for ewes being flushed for tupping. However, where available, fresh aftermaths (unsupplemented) can give lamb live-weight gains of up to 1 kg/week and finish short-keep lambs in 6–8 weeks. Table 6.1 illustrates the crucial importance of the 'D' value of the herbage. The higher the digestibility, the greater will be the intake of forage dry matter (DM).

Rested and fertilised aftermaths in the autumn can have a 'D' value as high as 68. However, if the digestibility is below 64, it can be seen that supplementary feeding will be needed to ensure the gain of 1 kg/week which is normally the target with short-keep finishing.

Long-keep lambs can be kept on grass in favourable seasons until the turn of the year, gaining around 500 g/week with no supplements. However, grass growth declines sharply from mid October, and the

Table 6.1 Feed potential of grazed herbage for weaned lambs

Forage quality ('D' value)	*30 kg live-weight*		*40 kg live-weight*	
	Expected DM intake (kg/day)	*Growth potential* (g/day)	*Expected DM intake* (kg/day)	*Growth potential* (g/day)
70	1.25	230	1.60	230
68	1.09	200	1.39	200
66	1.00	175	1.28	175
64	0.89	140	1.14	140
62	0.83	115	1.07	115
60	0.78	80	1.00	80

Source: J. R. Hopkins, ADAS *Nutrition Chemistry*.

feeding of hay, roots or cereals may then be necessary as grass intake declines. Supplementary cereals may be needed at 100 g/day rising gradually to 500 g/day.

A final important warning. If at all possible, keep store lambs off any newly seeded leys to be grazed by the flock in the following season. Otherwise, the next lamb crop will be at risk from parasitic worm infestation after turn-out in the spring.

Forage catch crops

Stubble turnips, forage rape and fodder radish have two big attractions. Firstly, they occupy land for only a part of the year, allowing another crop to be grown in the same season. Secondly, they are capable of providing a big bulk of cheap forage within about 10 weeks of sowing.

Fodder radish is very fast-growing, but unfortunately becomes over-mature with great rapidity, accompanied by a steep fall in 'D' value and palatability. The authors have found it to be almost impossible to manage.

Stubble turnips and rape are similar in their capacity to provide forage, mainly leaf, during two periods of the year when lamb keep can be short.

- Forage catch crops can be sown from mid April to early June to provide forage when grass growth is in the mid-season trough. Such crops can be direct-drilled into sprayed grass following a cut for silage. Direct drilling has been successful on many soils because it conserves soil moisture and may limit poaching of the land.

- Those sown from late June to mid August follow early-harvested winter barley, grassland, early potatoes or vining peas. The forage provided from September to the end of the year will finish short-keep lambs (with a concentrate supplement) or hold densely stocked long-keep lambs until the commencement of their finishing period at the turn of the year.

The introduction of stubble turnips into the UK in the 1960s was an important innovation for the flockmaster, and this crop usually out-yields forage rape (see Table 6.4). It is normal to graze turnips from about 10 weeks after sowing. They remain palatable for approximately 2–3 months, but are not winter-hardy and cannot be relied upon after early January at the latest.

Forage rape grows a little slower than stubble turnips, and lambs should be introduced to this crop gradually because of the danger of 'rape poisoning'. This danger can be reduced by sowing rape with a companion crop such as Italian ryegrass or turnips. Further worthwhile safety precautions are to remove the lambs from the rape overnight and to have hay on offer in racks at all times.

Root crops

In many arable areas of the UK, whole-season crops of turnips and swedes are grown to be folded by store lambs in the autumn and winter. Traditional yellow-fleshed turnips are sown in May or June and grazed from October to December. They grow more slowly than stubble turnips and most of their feed value is in the root. Lambs move from turnips to swedes in December, and may remain on the latter winter-hardy crop throughout the winter. The swede crop is capable of very high total yields of up to 100 t/ha; however, modern varieties tend to be of high DM content and are therefore quite hard; this has the twin effects of accelerating tooth loss and of reducing feed intake in lambs changing their teeth. Consequently, interest has increased in an alternative method of utilisation of the swede crop, namely harvesting, storage and feeding in the chopped form. In many trials this has resulted in increased intake and enhanced performance when compared to folding.

Mangolds and fodder beet are alternative root crops for harvesting and feeding either whole or chopped after storage. In the majority of areas they cannot be grazed through the winter because they are not winter-hardy. However, except in the wetter and hill areas they normally outyield the swede crop. At Rosemaund EHF, the DM yield per hectare of fodder beet has doubled that of swedes over the last 15

Plate 6.2 Lambs folded on roots

years. Flockmasters looking for a high yield of roots should consider fodder beet. This crop can be harvested easily (by a swede or sugar-beet harvester), stored with little loss in a straw-covered clamp and form part of a high-energy palatable ration.

At Rosemaund EHF, store lambs kept indoors and fed ad lib chopped fodder beet have gained in weight 40 per cent faster than similar lambs fed ad lib chopped swedes over a 70-day finishing period. The supplement fed to the lambs offered swedes was 250 g/day of mineralised rolled barley, and lambs on fodder beet had 200 g barley plus 50 g soya bean meal per day. Fodder beet can be fed whole or chopped to store lambs from November to May and its superiority over other forage crops in terms of DM yield and energy content can be seen from Table 6.4.

Kale and cabbages

The growing and folding of these leafy brassicas was studied in the 1970s by ADAS in the Beverley area of Yorkshire and at the nearby High Mowthorpe EHF. Although the DM yield per hectare of cabbages is no improvement on that of the swede crop they can be folded with less crop wastage, particularly when the lambs are changing

Table 6.2 Comparative yields of cabbages and root crops

Utilisation period	*November*		*December*		*January/February*	
Crop	*Turnip*	*Cabbage*	*Swede*	*Cabbage*	*Swede*	*Cabbage*
DM yield (t/ha)	6.2	7.4	8.5	7.4	8.8	5.7
Lamb grazing days (1000/ha)	4.6	5.2	8.3	5.1	11.1	6.3

Source: High Mowthorpe EHF.

Table 6.3 Comparative performance of lambs grazing cabbages and root crops

	Initial weight (kg)	*Final weight* (kg)	*LWG/week* (kg)	*Compound fed/lamb* (kg)
Roots	34.0	40.3	0.42	14.0
Cabbage	33.8	44.8	0.89	4.0

Source: High Mowthorpe EHF.

teeth. The results from High Mowthorpe showed that lambs grazing cabbages had better live-weight gains and produced heavier carcasses than lambs on root crops. In addition, the need for supplementary feeding was reduced. The favourable results from cabbages have not always been repeated with kales. Lambs tend to break off kale leaves and trample them underfoot, so increasing crop wastage.

The High Mowthorpe trials started in 1975, and over a 4-year period, the crop yields and lamb performance on cabbage were compared with those on turnips in November and with those on swedes in December and again in January/February. Table 6.2 gives the 4-year mean yields and the lamb grazing days per hectare achieved on the crops. The 4-year mean lamb performance is shown in Table 6.3.

Reduced crop wastage and enhanced feed intake by lambs folded on cabbages accounted for their superior performance. However, these two advantages can be achieved from root crops such as swedes or fodder beet if they are harvested, chopped, and fed under a roof. There has been a move away from the folding of forage crops in recent years which has benefited soil structure, crop rotations and lamb performance.

Sugar-beet tops

Arable by-products such as sugar-beet tops and residues from crops of Brussels sprouts, cauliflowers and cabbages represent 'something for nothing' to the lamb finisher in terms of feed cost. The expenses involved in folding them are mainly those of supplementary feed, fencing and labour.

Sugar-beet tops are available for lamb finishing from October to December and provide forage well balanced for protein and energy. Hay or straw should be available to the lambs on a run-back area.

Comparison of forage crops

It is traditional in many areas to graze a number of forage crops in sequence. Lambs may be on grass and stubbles in September and

Table 6.4 Forage crops compared

	Sowing date	*Utilisation period*	*Total yield* (t/ha)	*DM content* (%)	*DM yield* (t/ha)	*'D' value*
Stubble turnips (summer keep)	April/May	July–Sept.	56	9	5.0	74
Stubble turnips (autumn keep)	July/Aug.	Sept.–Dec.	40	9	3.6	74
Forage rape (summer keep)	April/May	July–Sept.	38	12	4.5	72
Forage rape (autumn keep)	July/Aug.	Sept.–Nov.	30	12	3.6	72
Sugar-beet tops	April	Oct.–Dec.	28	16	4.5	62
Cabbage	April/May	Nov.–Feb.	64	12	7.6	77
Swedes	April/June	Nov.–Mar.	70	10	7.0	82
Fodder beet	April/May	Nov.–April	70	16	12.0	84 (root)

Sources: ADAS Booklet 2394, *Store Lamb Finishing*.
ADAS Nutritional Chemistry Department, Leeds, Booklet, *Feeding for Efficient Lamb Production*.

October and then graze turnips, rape or sugar-beet tops if available during the last 3 months of the year. From January, the choice lies between folded swedes or cabbages or an indoor finish on diets based on chopped roots, hay, silage or high-level compound feeding. The forage crops with most to recommend them are compared in Table 6.4.

The utilisation of forage crops

The folding of forage crops was revolutionised by the invention of the light-weight electrified plastic fencing net, and more recently Harry Ridley (of Chilbolton Down fame) has marketed the Ridley Wrapper (see Plate 6.3), which is extremely easy to erect and take down.

Plate 6.3 Harry Ridley and the Ridley Wrapper
(*Courtesy* Tony Cave-Penny)

A good crop of swedes on free-draining land will provide grazing for around 1000 lambs/ha; at a live-weight gain of 1 kg per lamb per week this adds up to 1000 kg gain/ha. Moreover, this performance may be achieved with low-level cereal supplementation at the rate of around 300 g/head/day over the final few weeks.

It is, however, undeniable that the folding of forage crops is associated with a high level of crop wastage. At High Mowthorpe EHF, this was estimated at 40 per cent of the swede crop (compared with only 5 per cent when swedes were mechanically harvested) and at Liscombe EHF, estimates of wastage when folding swedes varied from 10 to 65 per cent.

Although it is unrealistic to believe that we can eliminate this crop wastage, certain measures will limit the loss and promote better performance from folded lambs as follows.

- Grow the crop on free-draining fields.
- Always provide a dry lie-back area; this facilitates the *gradual* introduction of the lambs to the roots, and is a suitable site for the water point, troughs and hay racks.
- Graze the crop in blocks of about 3 ha. In EHF trials this method has meant no more crop wastage than with strip grazing and less wastage in wet weather. The lambs are more settled, and the labour requirement is reduced.
- Limit grazing mobs to 200 or so. Most flockmasters have found these easier to manage than larger groups.
- Graze forage crops during their period of peak palatability.

The folding of swedes in late winter may pose particular problems because it coincides with the change of teeth in store lambs. This is doubtless one of the reasons for the finding at High Mowthorpe and Rosemaund EHFs that the mechanical harvesting and chopping of roots improves the efficiency of their utilisation dramatically compared with folding. Lambs fed chopped roots have given improved live-weight gains and had a lower supplementary concentrate requirement.

At High Mowthorpe, Suffolk-cross lambs held in store condition throughout the autumn were finished on swedes with barley supplement. Three methods of feeding swedes were compared in 1975 and 1976 (see also Table 6.5):

- Folding.
- Feeding chopped swedes outside in an open concreted yard.
- Feeding chopped swedes in a sheep house.

Table 6.5 Comparison of three methods of swede feeding

	Swedes folded	*Swedes chopped*	
		Fed outside	*Fed inside*
Starting weight (kg)	32.4	32.6	32.6
Final weight (kg)	36.0	43.0	41.9
No. of days on treatment	72.7	64.7	72.5
Daily live-weight gain (g)	49.0	160.0	129.0
Barley supplement (g DM/day)	310.0	86.0	96.0

Source: High Mowthorpe EHF.

Hay and silage

Hay and silage are expensive feeds and therefore most suited to the feeding of long-stay lambs to be finished when the lamb price is high. However, they allow considerable flexibility in that store lambs may be housed late in the year and fed hay or silage with the supplementary feed manipulated to achieve the desired finishing date.

Baled hay

Despite its many drawbacks, baled hay is a very convenient feed that can be purchased at any time and stored without deterioration. Store lambs will eat from 0.5 to 1.0 kg/day and with a daily supplement of 300–500 g of a cereal/protein mix (depending on hay quality) a gain of 1 kg/week can be expected. Table 6.6 shows that only average- or good-quality hay will give this level of performance. If poor hay or straw is fed, the compound ration must be considerably increased to make good the deficiency in forage quality.

Silage

Experience at Rosemaund EHF has shown that to achieve a good performance from store lambs (a gain of over 1 kg/week with low level supplementary feeding) grass silage of the quality listed in Table 6.7 is required.

The importance of a good fermentation cannot be too strongly emphasised, since lambs will ignore silage of low palatability. High dry matter and energy contents are necessary to optimise energy intake and therefore minimise the need for supplementary cereals. Chop length is also important, with a fine chop leading to higher intake than long chop. At Rosemaund EHF, the feeding of unchopped silage has also meant more wastage with housed lambs tending to pull the long

Table 6.6 Hay and straw diets for finishing lambs

Forage quality ('D' value)	*120 g daily gain*		*160 g daily gain*	
	Forage DM/day (kg)	*Compound/ day* (kg)	*Forage DM/day* (kg)	*Compound/ day* (kg)
30 kg lambs				
Good hay (61)	0.79	0.06	0.74	0.22
Average hay (57)	0.64	0.21	0.48	0.44
Poor hay (51)	0.50	0.35	0.47	0.49
Straw (45)	0.41	0.45	0.39	0.58
40 kg lambs				
Good hay (61)	1.01	0.08	0.94	0.30
Average hay (57)	0.83	0.22	0.76	0.48
Poor hay (51)	0.64	0.47	0.60	0.65
Straw (45)	0.53	0.59	0.49	0.76

NB: The compound feed is assumed to contain 88 per cent DM and ME of 12 (MJ/kg DM). The protein content of the diet DM should be at least 14 per cent.
Source: J. R. Hopkins, ADAS *Nutrition Chemistry*.

Table 6.7 Rosemaund EHF average silage analysis

DM (%)	25.0
pH	4.0
% in DM	
crude protein	17.0
MAD fibre	32.0
Ammonia CP as % total CP	10.0
Estimated 'D' value	66.0
Estimated ME (MJ/kg DM)	10.5
DCP (g/kg)	125
Ash content	<10
Fermentation	Lactic and stable
Chop length	Approximately 20 mm (precision chop)

NB: MAD = Modified acid detergent.
DCP = Digestible crude protein.
Source: Rosemaund EHF.

forage through the feed barrier and drop it in the bedding. This in turn increased the incidence of foot rot.

Lambs do not have an immediate taste for silage. It should therefore be introduced gradually, and 250–300 mm trough frontage should be allowed per lamb. If silage is fed only once, daily intake is encouraged by picking up any spillage and 'fluffing up' the forage in the troughs at the other end of the day. Uneaten silage must not be allowed to accumulate. Daily trough cleaning is essential.

Silage supplementation

Silage supplementation has been studied in numerous trials, and the authors are indebted to M. Appleton of ADAS, Liscombe EHF, who summarised the evidence from Liscombe together with Rosemaund and High Mowthorpe EHFs.

It is clear that all silages require supplementation if lamb live-weight gain is to be consistently in excess of 1 kg/week. Silage intake of store lambs normally lies between 2–3 kg/day (0.5–0.8 kg DM/day) and average-quality silage fed alone has given a live-weight gain of only 300–400 g/week.

Energy supplementation is frequently in the form of barley or oats, and trials have shown that whole and crushed cereals give similar results. It is known that supplementary cereal feeding reduces silage DM intake. This effect is negligible up to a cereal level of 300 g/day, but becomes highly significant when over 500 g/day are fed.

There is evidence that a relatively small inclusion of undegradable protein in the form of fish meal or soya bean meal may (unlike barley) enhance silage intake and also improve lamb performance. This can occur even when the silage is of high quality, as in a 1985/86 trial at Liscombe EHF. Suffolk-cross store lambs were fed 70 'D' grass silage to appetite alone or with 200 g/day barley or with 200 g barley plus 30 g fish meal/day (Table 6.8).

Table 6.8 Lamb performance on three rations

	Silage only	*Silage plus barley*	*Silage plus barley plus fish meal*
Live-weight gain over 56 days (g/day)	100	110	150

Source: Liscombe EHF.

With excellent silage on offer, the response to barley feeding in this trial was very small. However, there was a good response to the addition of fish meal at 30 g/day with the lambs finished at slightly heavier weights than on silage or silage plus barley. Most silages of a quality high enough to be fed to lambs require supplementary feeding of between 200 and 400 g/day to ensure a live-weight gain of around 1 kg/week. Appropriate compound feeds would be 90 parts whole barley : 10 parts fish meal or 85 parts whole barley : 15 parts extracted soya bean meal. A sheep mineral/vitamin supplement should be included in the concentrate at 2.5 per cent. The sprinkling of minerals on top of silage in the trough may limit silage intake and is therefore not recommended.

Finally, mention should be made of forage maize silage. In the southern areas of the UK where maize consistently yields well, it produces a big bulk of high-energy forage, and provides a means of finishing lambs on arable farms in the absence of grass. However, maize silage requires additional protein supplementation, and experience of its use for the finishing of store lambs is limited. In trials over 2 years at Rosemaund EHF, store lambs did not appear to relish maize silage, and intake and performance were therefore disappointing.

Indoor finishing on concentrates

Indoor finishing on concentrates is an interesting opportunist enterprise first developed in detail at the University of Newcastle. The theory is that lambs kept cheaply until the turn of the year and also tail-end lambs may be profitably finished indoors on high-energy rations as an alternative to selling them on the store market. The aim is to achieve rapid growth (necessary because the feed conversion ratio is steadily declining) and sell on to the firm lamb markets of March and April. Therefore, the lambs should be housed in December at the earliest and finished within 8 weeks.

Lambs must be kept in well-ventilated houses on dry beds, and the best performance is attained when the number per pen is limited to 25. The trough frontage allowance should not be less than 300 mm/lamb.

At Rosemaund EHF, a simple roofed pole barn with an open front and an expanded metal floor which was built by farm labour, has been completely adequate.

Unless the lambs have already been on trough feed, a change-over ration is recommended for the first few weeks. For this purpose a 50 per cent whole oats : 50 per cent dried sugar-beet pulp mix has been successful. This should be introduced before housing and the ration should be gradually increased as shown in Table 6.9. The lambs

Plate 6.4 Lambs eating compound pellets from starter trough

Table 6.9 Change-over ration for lambs

Week	*Ration* (g/head/day)
1	100
2	200
3	300
4	400

may then be housed and the ration further increased to appetite, normally a little over 450 g/head/day at first. Results from Redesdale EHF suggest that the feeding of concentrates twice daily to appetite is preferable to leaving the compound in front of the lambs at all times. The twice-daily feeding prevents the build-up of stale food. A simple, high-energy diet containing around 13 per cent CP should gradually replace the change-over ration from housing. An example is 9 parts whole barley to 1 part soya bean pellets. To this should be added 2.5 per cent of a mineral/vitamin supplement designed for indoor sheep, containing no added copper and the correct balance of calcium, phosphorus and magnesium to avoid urinary calculi (stones in the urinary tract of male lambs). Clean water must be available at all times, but hay is not necessary after the change-over period if the lambs are bedded on straw.

This is a tightly controlled method of lamb finishing and the results are quite accurately predictable. Cross-bred lambs housed weighing 30 kg will consume 35–45 kg of concentrate over a 6–7 week finishing period to reach 38 kg at sale. The live-weight gain should be from 150 to 200 g/day, and the feed conversion ratio range from 5 : 1 to 8 : 1. There is evidence to show that lambs shorn in the autumn may respond with significant improvements in feed intake and performance.

In recent years, High Mowthorpe EHF has gained much experience of this finishing method. No difficulties have been found in producing 18 kg 3L carcasses from Mule or from Scottish Blackface lambs, although the live-weight gain of the latter is, of course, lower than that of bigger lambs, and an attempt to produce 24 kg carcasses from Suffolk x Mule lambs is currently being made.

The Health of Store Lambs

Disease problems may cause poor performance and mortality in store lambs. It is therefore essential that measures for their prevention and control are worked out in advance with the farm's veterinary surgeon. The diseases most commonly encountered are as follows.

- *Parasitic worms*. Newly purchased lambs and lambs about to be housed or put on clean grass should be drenched to control stomach and intestinal worms, lung worms, and fluke on farms where this is likely to occur.
- *Clostridial diseases*. Purchased lambs should always be protected against these killer diseases by vaccination.
- *Pasteurellosis*. This can cause high lamb losses. It is triggered by any form of stress on the lambs, and detailed advice from your vet is essential. Vaccines are available.
- *Foot rot*. This may be brought in with purchased store lambs. It depresses performance because of the discomfort it causes. There is no doubt that foot rot is easier to eradicate when the lambs are housed on slats or on an expanded metal floor than when they are in a bedded house.

- *Orf*. Still an occasional problem, particularly in housed sheep; the desirability of vaccination should be discussed with your vet.
- *Coccidiosis*. Not common in lambs outside and store lambs will probably have developed immunity. The disease is more common in housed lambs and where necessary the use of a coccidiostat should be discussed with your vet.

The authors have found that disease problems can be kept to a minimum by thinking ahead about the management of the lambs. Prior planning should eliminate any abrupt changes in feeding and management which are the cause of so many health problems. With this attitude and detailed veterinary advice *in advance of trouble*, mortality in store lambs should not exceed 2 per cent.

The Economics of Lamb Finishing

The foundation for a profit is laid on the day of lamb purchase. A judgement must be made at this time on the likely value of finished lambs in the anticipated slaughter period. If the costs of production per lamb (forage plus concentrates plus vet and medicines plus miscellaneous outgoings) are then subtracted from the likely lamb sale figure, the affordable price of store lambs becomes clear. This must be low enough to allow all variable costs to be covered and the target gross margin per lamb to be realised (Table 6.10).

Table 6.10 Relationship between sale price and purchase price of lamb

Enterprise	*Approximate feeding period*	*Mortality*	*Percentage increase over purchase price*		*Target gross margin*
	(weeks)	(%)	to break even	to leave acceptable gross margin	(£/head)
Short-keep grass finishing	8	3	8	12	3
Grass and forage crops	12	5	15	36	6
Long-keep roots	20	5	27	70	9

NB: The gross margin is the feeder's margin (difference between purchase price and sale price of the lamb) minus all the variable production costs previously mentioned.
Source: ADAS Booklet 2394, *Store Lamb Finishing*.

Table 6.11 Overall results from lamb-finishing enterprises 1986/7

		Average (£/lamb)		*Top third* (£/lamb)
Lamb sales		42.55		46.55
Less lamb purchase price		33.80		33.72
Output (feeder's margin)		8.75		12.83
Variable costs				
purchased feed	2.18		1.62	
grass	0.23		0.21	
forage	0.57		0.77	
Total feed and forage	2.98		2.60	
vet and medicines	0.23		0.26	
other costs	0.29		0.30	
Total variable costs		3.50		3.16
Gross margin/lamb		5.25		9.67
Physical performance				
average group size		459		441
average number of days feeding		85		108
feed cost/lamb/week (£)		0.25		0.19
mortality (%)		1.8		0.9

Source: MLC.

It is impossible to include financial budgets to cover the range of store-lamb finishing systems. Table 6.11 gives the average financial and physical performance of 71 store-lamb finishing enterprises recorded by the MLC in 1986/7, and also the performance of the top-third units. These figures are not specific to any particular finishing method. However, they do show the effects of the various aspects of physical performance on the financial result, and provide a target with which the flockmaster can compare figures from his own finishing enterprise.

The top-third flocks kept the lambs 23 days longer than the average, and sold when the guide prices were higher. They achieved a £4/lamb higher sale price, and as their lamb purchase price was similar, the top-third producers had a feeder's margin which was £4/lamb higher than the average. They also spent less on purchased feed, and showed a big increase (£4.42/lamb) over the average GM.

Chapter 7

Lambing Out of Season

Most home-producers lamb their flocks at the 'natural' time – in the spring of the year – selling as many lambs as possible off grass from June to October. Store lambs (home-bred or purchased) are then finished in the late autumn and winter. This has left a gap in the market in the first half of the year, traditionally filled by imports from overseas producers, particularly those in New Zealand.

The diversification policy of the New Zealand government has resulted in much more of their lamb now being sold on newly found markets throughout the world, and less coming to the UK. This has stimulated interest in sheep systems which market prime-quality home-produced lamb in the period November – May, thereby ensuring continuity of supply for the British housewife. Further impetus to this trend has come from the seasonal price structure for lamb under the present UK guide prices.

The finishing of store lambs is one method of supplying the market in the period from November, and this is fully discussed in Chapter 6. The alternative is to lamb the flock outside the normal breeding season. The latter may be defined for most lowland ewes as from February to April. Lambing outside this 'natural' period can be aided by choice of breed, the use of vasectomised rams and by one of several hormonal treatments.

The three important out-of-season breeding systems are detailed in this chapter: early lambing, late lambing and frequent lambing.

Early Lambing

Marketing

The first essential when considering setting up an early lambing flock is the availability of a suitable market. Early born lambs sold young tend to lose more weight during travelling to market than older lambs. Hence, it is preferable to seek local outlets. It is important also to clarify weight and fat class requirements, since these may differ from market requirements later in the season.

Buyers are understandably reluctant to purchase early spring lamb unless a continuous supply can be ensured. With this requirement in mind, producers should consider forming marketing groups to supply early lamb on a regular basis. Such groups are already operating successfully in Scotland, Northern Ireland, Wales, the Midlands and south west England, establishing a profitable trade for early lamb.

Breed type

The ewe is a seasonal breeder but some breeds have a more extended breeding season than others and there is a wide variation within breeds.

The onset of oestrus is triggered by shortening day length and it can be seen from Table 7.1 that some breeds come into oestrus earlier in

Table 7.1 The effect of breed on duration of the breeding season

Breed	*Date of onset of oestrus*	*Date of mid breeding season*	*Date of end of oestrus*
Dorset Horn	15/6–21/8	12/11	22/1–22/4
Suffolk	12/9–23/10	25/12	7/2–20/4
Romney Marsh	16/9–19/10	18/12	20/1–23/3
Border Leicester	24/9–13/10	10/12	18/12–11/3
Scottish Black Face	26/9–10/11	23/12	17/1–3/4
Welsh Mountain	15/10–11/11	22/12	1/2–25/2

Source: Hafaz (1952).

the summer/autumn than others. These breeds are the most suitable for an early breeding flock.

The Dorset Horn is the only native breed which will consistently conceive outside the normal September – January breeding season with no hormonal assistance. Their ability is mainly used to advance the date of mating and Polled Dorset crosses are particularly valued in early lambing flocks. The Suffolk breed may also commence its breeding cycle in the autumn earlier than most other breeds and Suffolk cross ewes are widely used in early lambing flocks for this reason.

The ram effect

After the initial introduction of the rams at the beginning of the breeding season, ewes come on heat in two distinct peaks. These occur at days 18–20 and at days 24–26 after the introduction of the rams.

The reason for this is that within a few minutes of close proximity to the ram, his scent or pheromones stimulate the ewe's pituitary gland to produce hormones. These hormones induce the Graafian follicles in the ovary to mature and shed eggs (ovulation).

Ovulation by this means is unusual in two ways. Firstly, it is not accompanied by the usual sexual behaviour which indicates that the ewe is on heat; it is known as a 'silent heat'. Secondly, in about half of the ewes the corpora lutea (yellow bodies) in the ovaries, which usually function for 16–17 days, recede within a few days. A second ovulation, also accompanied by a silent heat, occurs in a further 6–7 days after the first silent heat. The corpus luteum then develops for the normal 16–17 days before receding and an ovulation occurs which is accompanied by normal signs of heat.

Progesterone is produced by the corpus luteum and progesterone priming is essential for a normal heat period and successful implantation of a fertilised egg.

Teaser rams

In order to bring the timing of the first normal heat forward, teaser rams can be used. A teaser ram is an entire ram which has been vasectomised to render him infertile. However, unlike castrated males, he still produces the male pheromones which stimulate the ewe's pituitary gland.

Teasers are used most effectively at the beginning of the breeding season when the ewes have been kept well away from the rams (entire or teasers) for a month before the teasers are introduced. The teasers are run with the ewes from 17 days before the planned date of commencement of tupping. After 15–16 days, the teasers are removed and replaced by entire rams; this will coincide with the ewes' first true heat and most should be successfully mated within the first week.

To summarise, teasers can be used both to advance the normal breeding season by 3–4 weeks (but no more) and to synchronise the ewes at mating so that the lambing period is more compact.

One teaser ram per 100 ewes is sufficient. There is some evidence that Dorset and Finn × Dorset Horn vasectomised tups are among the most effective, but any active ram can be used. It should be remembered that teasers will bring about a partial synchronisation of oestrus only, being much less effective than progesterone-impregnated sponges. However, they have a beneficial effect on the management of ewes bred early out-of-season and this is achieved at a low cost. The above sequence of events is illustrated in Figure 7.1.

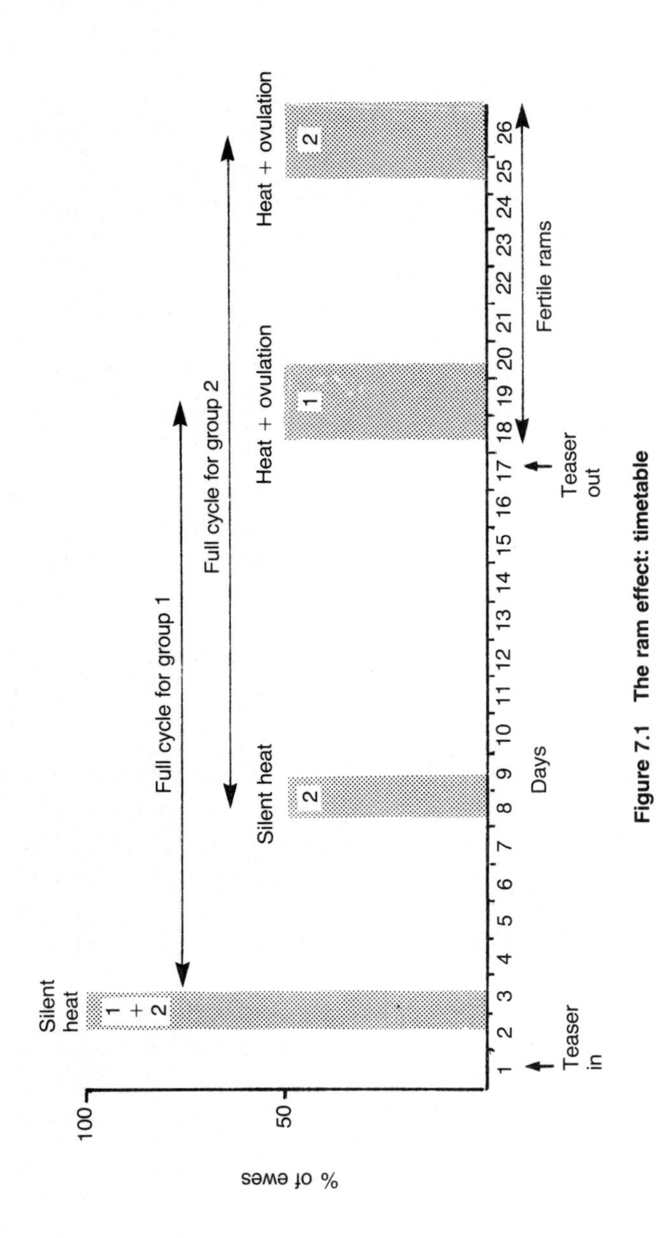

Figure 7.1 The ram effect: timetable

The progesterone sponge and pregnant mare serum gonadotrophin (PMSG)

The use of the progesterone-impregnated intra-vaginal sponge together with pregnant mare serum gonadotrophin (PMSG) is a convenient method of inducing an early normal heat in ewes in an early lambing flock. The manufacturers' instructions should be followed. The degree of synchronisation is much greater following the use of a sponge than when relying solely on teaser rams, and the PMSG ensures ovulation. The sponge releases a controlled amount of progesterone which is normally produced by the corpus luteum.

The sponge is carefully inserted into the vagina (using the instrument provided) so that it lies against the cervix. When the sponge is in place all that can be seen are two threads protruding from the vulva. A veterinary surgeon should demonstrate this technique for the first time, but most shepherds are well able to perform the task if care is taken and careful hygiene observed.

Normally, the sponge is removed after 12–14 days by gently pulling the threads. Very occasionally, the threads will pull out from the sponge and if the sponge cannot be removed by a clean, lubricated finger the ewe should be taken to the vet who will remove it using forceps.

When the sponge is removed it is accompanied by a small amount of evil-smelling mucus but this has no adverse effect on conception rate. A recommendation from the Rowett Research Institute is to dust the sponge with an antibiotic before insertion rather than to use the antiseptic cream sometimes supplied with the sponges. This not only eliminates the mucus discharge but fewer sponges appear to be lost during the 12–14 day period. Practical experience at Rosemaund EHF has shown that 2–4 per cent of the sponges can be lost and that all those ewes which had lost their sponges came on heat with the other ewes.

Immediately after sponge removal the ewe is injected with PMSG. This has the same effect on the ovaries as the hormones normally released by the pituitary gland. These have the effect of causing the follicles in the ovary to mature and rupture (ovulation).

The dose rate of PMSG is measured in international units (iu). It is desirable to use a dose rate high enough to stimulate ovulation but not so high that many eggs are shed (super ovulation).

Super ovulation is undesirable on two counts. Firstly, if too many eggs are released and fertilised, none of them may be successfully implanted in the uterus. Secondly, if the eggs are implanted, the ewes will give birth to three or more lambs. The birth of small lambs causes rearing problems, and such multiple-birth lambs are not likely to be finished in time for the Easter trade.

The dose rate of PMSG will depend on the breed of the ewe and on how early in the season mating is to take place. The normal dose rate varies between 300 and 500 iu for Suffolk cross ewes lambing in early January.

The rams should not be introduced to the ewes until 48 hours after sponge removal even though it is obvious that some ewes are on heat within 24 hours. The main reason for this is that progesterone is slightly spermicidal and the delay of 48 hours will allow it to disperse from the vagina. A further reason for the delay is that it will prevent the rams repeatedly serving the few ewes coming on heat first and so depleting their sperm reserves. At Rosemaund EHF it has been observed that, following the use of sponges, all the ewes are served within 24 hours after the rams are turned in. Most are mated within the first 12 hours. This means that the ratio of rams to ewes used on a synchronised flock must be higher than normal. It is generally recommended that a ratio of 1 ram : 10 ewes is needed. To reduce the amount of walking and seeking a ram must do, the ewes should be confined to a small area stocked at 50 ewes/ha.

The Suffolk rams used at Rosemaund EHF are all raddled with the same colour when they are turned in with the ewes. When all the ewes have been covered, the rams are removed with the exception of 1 ram : 50 ewes which remains (with a different raddle colour) to serve any ewes which failed to hold to the first service. These rams are removed after a further 20 days and the ear numbers of the ewes which failed to hold to the first service are noted.

The use of the progesterone sponge and PMSG injection (both are needed in most cases for early breeding) can advance the date of first oestrus and ovulation by up to 6 weeks. The ewes must be fit (body score 3–3.5) and the tups fit and sexually active. Results vary with the breed of ewe but experience indicates that 50–80 per cent of ewes conceive to first service in mid August following the use of sponges and PMSG. The disadvantages are the treatment costs and the large number of tups required. Housing is needed in many areas and a generous availability of staff to avoid mismothering after lambing.

Melatonin

It is a well-known fact that under natural conditions ewes start coming into oestrus in the autumn so that they lamb down in the spring. The controlling mechanism for this has only recently been explained. During hours of darkness, melatonin is produced in the pineal gland in the brain. In the short summer nights, melatonin levels do not stay high long enough to trigger a heat period but as the nights lengthen during the autumn, levels remain high for longer periods and the oestrus cycle begins.

Natural melatonin has been recovered from the process of decaffeinating coffee beans. Following early research in Australia, various researchers have administered melatonin to ewes during the summer months in attempts to induce early oestrus.

In early experiments, the melatonin was fed with rolled cereal grains in the late afternoon (this was an attempt to elevate melatonin levels in the blood prior to the evening rise in naturally produced melatonin).This method induced early oestrus but in practice proved difficult and expensive to operate. Consequently, alternative methods of melatonin administration were sought based on implants or boluses.

Implants containing melatonin have been developed for injection in either the ear or the neck. Alternatively, a soluble glass bolus containing melatonin can be administered orally with a balling gun. The bolus lodges in the reticulum (second stomach). The implants and the bolus gradually dissolve over a period of approximately 60 days and slowly release melatonin.

At Rosemaund EHF, only the soluble glass bolus has been tested and the results have been very encouraging. The boluses were given to Mule ewes 42 days before the introduction of the rams in mid August. In an experiment carried out over 2 years, all of the 80 ewes which received the melatonin were served and conceived during the first 18 days after the introduction of the rams. In the control ewes which did not receive melatonin, only 27 per cent of the ewes came into oestrus. This resulted in a lambing percentage (number of live lambs born per 100 ewes put to the ram) of 162 from treated ewes compared with a lambing percentage of 45 from the control ewes. This was regarded as convincing evidence of the effectiveness of melatonin in bringing forward the breeding season in an early lambing flock.

The use of melatonin does not apparently increase the litter size significantly compared with the use of sponges and PMSG. This can be an advantage if the use of melatonin does not result in high numbers of triplet and quadruplet lambs.

The Rosemaund EHF results on the use of melatonin are similar in most respects to those reported by Dr Haresign of Nottingham University and his colleague Dr Williams using melatonin ear implants, and by the University of Surrey researchers in co-operation with Pilkington Glass using the bolus.

In the summer of 1987, a total of 1450 ewes on 14 sites in the UK were treated with the bolus in trials supervised by ADAS or the Scottish agricultural colleges. The boluses were given 42 days before the rams were introduced, and 95 per cent of the ewes conceived to their first service to lamb between Christmas and the end of January. In these on-farm trials the melatonin-treated ewes produced 20 per cent more lambs per ewe than untreated control ewes in the same flocks.

If we accept that the early lambing flock normally has a lambing performance approximately 15 per cent poorer than the spring lambing flock, the use of melatonin may eliminate this disadvantage. This would make December/January lambing an attractive proposition for flockmasters with available winter feed and housing.

The impending availability to flockmasters of melatonin will be an important breakthrough, particularly since it is a natural product which cannot harm sheep or humans.

The inducement of lambing

Despite the fact that the majority of sponged and injected ewes are served on the same day, they lamb down over a period of 1 week. To tighten the lambing period further, the ewes can be induced to lamb down over a 48 hour period. However, good record-keeping is essential to avoid inducing ewes which returned to the first service and held to the second one 17 days later.

Mule ewes at Rosemaund EHF have been induced by injecting with a cortico steroid, (16 mg betamethazone) on day 144 after mating. Lambing has usually started 36 hours after the injection and has been completed within 24 hours. Occasionally, 1–2 per cent of the ewes did not respond to the injection and lambed down 1 week later.

With the degree of accuracy which results from this inducement technique, it is possible to manage the ewes so that they lamb down on a predicted date, and this has advantages particularly for a small farmer who relies on part-time assistance, since he or she can plan to lamb down at the weekend when help is available. Conversely, lambing can be planned to take place in the middle of the week, thereby avoiding weekends, which can enhance labour availability and/or reduce labour costs.

Flockmasters wishing to manage an early lambing flock in this way should first decide the date on which they wish to start lambing and then work back using Table 7.2.

It must be remembered that the gestation period of ewes varies by several days depending on breed. At Rosemaund EHF, Mules have averaged 146–7 days and Finn × Dorsets, 142 days. In mixed-breed flocks with ewes of varying gestation periods, the practice of inducing lambing is not recommended. However, in single-breed flocks where tupping dates are accurately recorded, the practice can greatly improve ease of management at lambing time.

Table 7.2 Breeding time-table using intra-vaginal sponges and induced lambing

Day	*Operation*
−14	Sponges in
−2	Sponges out and PMSG injected
0	Tups in
12–14	Tups re-raddled
20	Tups out
144	Induce
146	Lambing starts
147	Lambing over

Source: Rosemaund EHF.

The stimulation of an early synchronised heat and ovulation in ewes, and the inducement of lambing by hormonal treatments is not achieved without increased costs. For the sponge, PMSG injection and inducing injection, the total cost in 1988 ranges from £4 to £6/ewe.

The use of the sponge and the PMSG injection is effective in promoting an acceptable lambing percentage in ewes to be lambed before mid January. The cortico-steroid injection to induce lambing is an optional extra.

Early lambing means lambing-down the flock in the October–January period and there are several reasons why lambing so early in the breeding season is attractive.

- Lambs may be sold at high prices at Easter.
- The lambs are usually reared indoors so the ewe stocking rate is high on a yearly basis.
- Lambing takes place at a time when labour requirements on other enterprises, especially on arable farms, are low.
- Older ewes can be used if lambed indoors.
- The splitting of flocks so that a number of ewes lamb early and a number at the normal time spreads the labour requirement and also reduces the number of rams required.
- Although the gross margin per ewe may not compare favourably with that of the spring-lambing ewe, higher stocking rates and, consequently, a higher gross margin per hectare can be achieved.

There are also disadvantages.

- The provision of sheep housing is likely to be necessary in most areas.
- Feed costs are likely to be higher for both ewes and lambs.
- Early lambing normally results in lower prolificacy.

Early lambing may be achieved by outdoor flocks in mild areas of the country (traditional early lamb production) or, more commonly, by winter-housed flocks.

Traditional early lamb production

Traditional early lamb production has been practised successfully for many years in the south west of England. The breed of ewe used is the Dorset Horn or Polled Dorset, but during the last few years some Finnish Landrace crosses have been introduced to increase prolificacy.

The early lambing ewes are flushed on grass during the last 2 weeks of May and tupping takes place in late May and early June. The pregnant ewes are kept cheaply at grass (with minimal concentrate feeding) until they lamb in October. The ewes usually lamb out of doors and after lambing the flock may be subdivided into ewes rearing singles and those rearing twin lambs. Only ewes rearing twins are fed concentrates and only twin lambs receive a creep ration. When the autumn grass is grazed down, the flock is folded on catch crops such as rape or stubble turnips and then on swedes which carry the flock from December/January until late March. The lambs are sold from February to May. The only change that has occurred in recent years is that electric fences have replaced the traditional wooden hurdles.

The system depends for success on the extended breeding season of the Dorset ewes and on mild snow-free winters and free-draining land. This outdoor system of early lamb production builds up soil fertility on the lighter chalky soils.

Housed systems for early lamb production

The ewes are normally lambed in December or in January. As this is in advance of the normal breeding season for most ewes, either teaser rams or hormone treatments (or both) are used. Melatonin is an alternative which will shortly be available.

Housing requirements

Well-ventilated, draught-free accommodation is essential with plenty of clean dry bedding for the ewes and lambs. Dirty bedding increases the risk of coccidiosis. Both ewes and lambs should have access to an ad lib supply of clean water. At Rosemaund EHF, small individual water bowls are preferred to a communal trough because they are easier to clean and to thaw out in frosty weather (Table 7.3).

Table 7.3 Space recommendations for Mule ewes and their lambs

Ewes before lambing	1.1 m^2/ewe
Ewe plus 2 lambs	2.0 m^2/ewe
Creep feed area	0.3 m^2/lamb
Lamb area after weaning	0.5 m^2/lamb
Creep feeder trough length	64 mm/lamb

Source: Rosemaund EHF.

Table 7.4 Compound fed to ewes at grass in December

	(kg/t)
Whole barley	775
Extracted soya bean meal	200
Mineral/vitamin supplement	25

Source: Rosemaund EHF.

Ewe nutrition

The nutritional requirements of pregnant ewes in the early lambing flock are the same as those of ewes lambing at the normal time (discussed in Chapter 4). In practice, the January-lambing ewes at Rosemaund EHF are at grass during the first weeks of December on fairly sparse grazing. To supplement the grass, they are fed a concentrate (Table 7.4) at 250 g/ewe/day during the penultimate week at grass and at 500 g/ewe/day during the final week at grass. The compound feed contains 16 per cent crude protein (CP) and has a metabolisable energy (ME) of 12.3 MJ.

The ewes are housed during the last 2 weeks of December and the amount of compound fed varies with the quality of roughage offered. If good-quality silage is available (ME over 10.8 MJ) the concentrate feeding level remains at 500 g/ewe/day. If the silage has an ME below 10.8, the amount of concentrate fed is increased gradually to as much as 700 g/ewe/day just before lambing. After lambing, the ewes are fed ad lib silage supplemented with 800 g/ewe/day of the same 16 per cent CP compound fed before lambing. From 1 week before weaning, which is when the lambs are 6 weeks old, the soya bean meal is omitted from the ration and this has the effect of reducing the milk yield. This is beneficial both to the ewe and to the lambs. It helps to prevent the ewes developing post-weaning mastitis and it encourages the lambs to eat more of the creep feed.

Plate 7.1 Early lambed ewe eating high-quality hay

Plate 7.2 Swing-over trough for feeding silage to ewes and lambs, left trough in filling position, right in position for ewes to eat

The feeding of silage in a barrier-type trough has proved ideal before lambing but not very convenient after lambing. The young lambs climb through the bars and lie on the warm silage. After urinating on the silage they escape from the pens and gambol in the passageway. Subsequently, some lambs return to the wrong pens and are, of course, unable to find their mothers. To solve this problem of lambs escaping, swing-over troughs were installed at the front of the pens (Plate 7.2). Silage is delivered into the trough from a forage box and the shepherd then swings the trough over 60° so that the ewes can eat the silage but the lambs cannot escape.

In 1987/8 the Rosemaund EHF early-lambing ewes were fed straw as the only roughage from 8 weeks before lambing to weaning. Before lambing, the compound contained 16 per cent CP and was allowed at a flat rate of 845 g/ewe/day. After lambing, the CP content of the compound was increased to 20 per cent and it was fed at 1.4 kg/ewe/day for 4 weeks. Subsequently, the ration was gradually reduced to a level of 500 g/ewe/day by weaning time.

This was the first winter in which straw had been fed at Rosemaund EHF to early-lambing lactating ewes and they produced healthy lambs weaned at good weights. The single lambs gained 310 g/day and twin lambs 280 g/day prior to weaning.

Lambing organisation

It is essential that sufficient individual lambing pens are available. At least one pen per three ewes is recommended where ewes are synchronised. Ewes with singles, providing they are housed, need not go into an individual pen. Ewes with good healthy twins need only stay in the pens until a family bond is formed (normally within 48 hours) and ewes with triplets or those with weak lambs must stay in the pens for longer periods.

One experienced shepherd can lamb down 100 ewes in 2 days with two provisos: firstly, that there is always someone on hand to take a ewe requiring veterinary assistance to the surgery; secondly, that assistance may be needed with a difficult lambing if the shepherd's time is committed to the supervision of other ewes lambing simultaneously.

Rearing the lambs

The lambs should be offered a proprietary 18 per cent CP creep feed from 10 days old. At Rosemaund EHF, this has been fed in a metal sheep trough which had weld mesh fitted over the top (Plate 7.1). This prevents the young lambs standing in the trough but allows them to

Plate 7.3 Early weaned lambs and creep feeder

reach the creep feed through the mesh. For the first week, the creep pellets are changed three times a day. The stale pellets are thrown to the ewes and replaced by fresh ones. It is important to keep the creep feed appetising to encourage young lambs to eat.

If the creep feed contains nutritionally improved straw, no other roughage need be fed. From about 3 weeks to slaughter the creep feed is best fed in a specialised creep feeder. At Rosemaund EHF, the lambs eat an average of 60 kg of creep feed from birth to slaughter at about 36 kg live-weight.

This system of early weaning December/January-born lambs at 6–7 weeks old and finishing them indoors has given consistently good results. The alternative – of turning ewes and lambs out to grass in the early spring – is cheaper but less attractive if the target sale date is Easter.

The early lambing system is financially assessed in Chapter 13.

Late Lambing

Lambing ewes late in the season, in the month of May, and selling the lambs from November to February is another method of producing lambs out of season. Unlike the early lambing systems already discussed – which rely on expensive housing and food – this system may be based on cheap feed provided by grazing and home-grown forage crops.

Liscombe EHF – an ADAS upland farm on Exmoor – has led the way in developing the late-lambing system. At Liscombe, Welsh Halfbred ewes are put to the ram in mid December–mid January with no supplementary feeding at grass. The flock is stocked until mid April at 10 ewes/ha reducing to 5 ewes/ha. During this period, hay is provided and the average ewe intake is 150 g/day. Feed blocks have been provided during February and March and 3 kg/ewe eaten during this period. In mid April the ewes are moved to good fresh grass which they graze without supplementary feeding until they are housed in the second week of May, when lambing commences. After 2 days inside, the ewes and lambs are turned out on to parasite-clean grassland at 20 ewes/ha and later on to silage aftermaths. Again, no supplementary feeds are offered to either ewes or lambs other than a magnesium block. Weaning takes place in the first week in August when the lambs are about 20 kg in weight. The newly weaned lambs are drenched against worms and grazed on silage aftermaths. The ewes continue to graze the original pastures at 19 ewes/ha.

In late October, the Liscombe EHF lambs are housed and finished on silage-based rations to be sold during the winter period. Meanwhile, the ewe flock is allowed to winter graze over the whole of the available grassland at 2 ewes/ha.

Mating the ewes late in their breeding season had a detrimental effect on lambing percentage; this averaged 130 per cent compared with a mid March lambing performance of about 160 per cent on the same farm.

However, lambing performance has been improved by injecting the ewes with Fecundin which restored the lambing percentage to 159. This has been economically worthwhile.

Late lambing systems have also been developed at the Institute for Grassland and Animal Production (IGAP), Hurley, Berkshire, and at Harper Adams Agricultural College. Since the ewes are mated towards the end of their natural breeding season, it is not surprising that researchers at Hurley (as at Liscombe EHF) reported a poorer lambing percentage than that normally expected from ewes lambing in March/April. There the performance of Masham ewes declined from 191 per cent in the spring to 178 per cent with May lambing.

This system involves low feed costs for both ewes and lambs, with concentrate feeding kept to a minimum. Lambs may be finished at a time when prices are high either indoors on silage or outdoors on cheaply grown forage crops.

Late lambing fits in best on mixed cattle and sheep farms. If cattle are housed in the autumn, this allows the ewe flock the run of the grassland at a low stocking density throughout the winter months.

Frequent Lambing

Frequent lambing means breeding the ewes more often than once every 12 months, and the objectives are as follows.

- To increase the profitability of the flocks.
- To offer prime young lamb for sale throughout the year.

When compared with other breeding farm animals, whether it be dairy cows, beef cows, pigs or poultry, the ewe does not appear to work very hard for her keep (Table 7.5). She may enjoy 2–3 months of the year resting from pregnancy and lactation and this 'holiday'

Table 7.5 Annual reproductive cycle of the dairy cow and the ewe

	Dairy cow (months)	*Ewe* (months)
Pregnant	9	5
Lactating	10	5
Pregnant and lactating	7	0
Resting	0	2

contrasts with the absence of a rest period for the dairy cow. Of course, the ewe has the advantage of multiple births but the frequency of reproduction is less than impressive, especially when we consider that, in theory, a ewe can breed every 6 or 7 months. In practice, this frequency has proved difficult to sustain on a flock basis, although some individual ewes will lamb twice in a 12-month period.

A more practical approach is to manage the ewes to lamb down every 8 months. Much of the research work on 8 months lambing has been pioneered by Dr John Robinson at the Rowett Research Institute and much excellent development work on the system carried out by Dr Mike Tempest at Harper Adams Agricultural College, where the type of ewe favoured for the system was the Finnish Landrace × Dorset Horn. The Finn was chosen for its prolificacy and the Dorset Horn for its extended breeding season.

At Rosemaund EHF, this specialised cross has been compared with Welsh Halfbred, Welsh Halfbred × Suffolk and Mule ewes for its suitability for 8 month lambing systems.

Rosemaund EHF flock management

In an 8 month frequent lambing system, ewes lamb in January, September and May. This means that every other year ewes lamb down in January and September of the same year. As this could disrupt the orderly management of the farm, it has been found more convenient to split the frequent lambing flock into two sub-flocks. These sub-flocks are managed in exactly the same manner but one flock is permanently 4 months behind the other so that the two flocks complement each other in producing lambs on the farm every 4 months (Table 7.6).

The 8 months breeding pattern is divided into three parts (Table 7.7) as follows:

- 1 month flushing.
- 5 months gestation.
- 2 months lactation.

Table 7.6 Lambing times

Flock A	*Flock B*
January	
	May
September	
	January
May	
	September

Table 7.7 Management of one flock through breeding cycle of 2 years

	Breeding sequence		
	1	*2*	*3*
Flushing	July	March	November
Tupping	August	April	December
Lambing	January	September	May
Weaning	February	October	June
Selling lambs	Easter	January/February	September

Source: Rosemaund EHF.

Ewe management

Such a tight schedule as that described above cannot be maintained without the use of hormones. The ewes have to be synchronised with progesterone sponges and injected with PMSG at all three mating periods. In August and December – months which are only just outside the normal mating period – the ewes responded to 500 iu PMSG. For the April mating, when all ewes are in deep anoestrus, 750 iu of PMSG had to be given.

January lambing

For January lambing, the flock management has been exactly the same as the early lambing flock in that the ewes were kept out at grass until 2 or 3 weeks before lambing was due. They were then housed and the lambs were reared indoors for the Easter market. Weaning took place when the lambs were 6–7 weeks old and ewes were taken away from the lambs and allowed only straw and water for 5 days. They were then either turned out to grass or fed silage indoors until spring conditions were suitable for turn-out. Care must be taken to ensure that the ewes do not lose too much condition because they have little time to regain it before mating again.

September lambing

The ewes were flushed in late March and early April on grass only. Mating took place in April and the ewes were kept out at grass until 2 weeks before they were due to lamb. A barley/fish meal ration was fed for the last 2 weeks of pregnancy at 250 g/ewe/day for 1 week and 500 g/ewe/day for the final week. In 1 year, the ewes were grazing particularly lush grass and the compound ration was not fed. This resulted in poor lamb birth weights, which indicated that the value of autumn grass may easily be overestimated.

The ewes lambed-down indoors and after 48 hours were turned-out to clean pasture with their lambs, where they continued to receive the barley/fish meal ration at 500 g/ewe/day for 6 weeks. At 6½ weeks after lambing, the flock was housed and the ewes taken away from the lambs at 7 weeks and fed straw only to dry them off. The lambs were finished indoors on a variety of feeds to meet rising market prices in the early weeks of the year.

May lambing

There has normally been insufficient high-quality grass in November to flush the ewes so they were fed 250 g/ewe/day of barley/fish meal ration for 3 weeks before and after mating. Usually they were housed soon after the tups were removed. One year there was a blizzard on the day the tups were turned in and this resulted in a poor lambing percentage; this is one of the hazards of mating a flock outside in December when the weather can be very variable. Where the ewes have been synchronised, the risk is especially high since wild conditions over the 2 days of mating can be quite disastrous.

The ewes have been lambed indoors and turned out to grass with their lambs 2 days later. At weaning in June, the flock has been housed for 3–4 days. The ewes are then removed and allowed only straw and water until dried-off and subsequently returned to grass. The lambs are finished indoors during the autumn and have therefore not achieved as high a market price as the lambs born in January or in September. Alternatively, they could be kept cheaply on forage crops (as at Liscombe EHF) for sale at a better price after the turn of the year.

Lambing performance

Results from the frequent lambing flock have been recorded for 3 years at Rosemaund EHF. There were wide variations in performance between the three lambing dates.

At the January lambing, the performance was quite predictable. The Mules and Suffolk crosses lambed down at just over 140 per cent, the Welsh Halfbreds at just under 120 per cent and Finn × Dorset ewes at 180–200 per cent.

May lambing usually resulted in high numbers of lambs being born, apart from the year previously mentioned when tupping took place in blizzard conditions in December. The Mule ewes lambed down at about 185 per cent and the Finn × Dorsets at 200 per cent.

The September lambings were always a disappointment. Most of the ewes were raddled at tupping in April but in some years, over half of the Mule and Suffolk cross ewes were barren, resulting in a lambing performance of between 66 and 81 per cent. The Finn × Dorset ewes performed better, with lambing percentages of about 130–150 per cent in most years.

The September lambing performance was not as disastrous as it might appear at first sight. The ewes were scanned in July to confirm pregnancy at about 80 days after mating and barren ewes were then taken out of their flock and joined their twin flock. They were tupped in August and lambed down the following January, 12 months after their previous lambing.

Table 7.8 Average number of lambs reared per ewe

Lambing date	*Mule ewes*	*Finn × Dorset ewes*
January	1.40	1.85
September	0.75	1.40
May	1.85	2.00
Lambs reared over a 2-year cycle	4.00	5.25
Lambs reared/annum	2.00	2.62

Source: Rosemaund EHF.

Lamb sales

Researchers at the North of Scotland Agricultural College studied frequent lambing under commercial conditions in the 1970s. They reported sales of 1.9 lambs/ewe/year from a flock of Finn × Dorset ewes and sales of 1.4 lambs/ewe/year from a Greyface flock.

The frequent lambing performance of the Rosemaund EHF flock is summarised in Table 7.8. Although the number of lambs reared per ewe per year was high, so were the feed and hormone costs. Gross margin per ewe from the Mule ewes was only a little higher than the margin from the early lambing flock, where ewes lambed down annually in December/January (see Chapter 13). Although the lambs from the Finn × Dorset ewes did not have such good conformation as those out of Mule ewes they, nevertheless, achieved the better margin.

In every year of the trial, the September lambing performance was poor and, despite experimentation with various levels of PMSG, the percentage of barren ewes remained unacceptably high. This inability to reduce the number of barren ewes at the April mating led to the abandonment of the development work on frequent lambing at Rosemaund EHF in 1986. However, the future availability of melatonin should increase the chances of success with out-of-season breeding, and the experiments may well be resumed. Meanwhile, frequent lambing systems remain attractive in theory but unproven in practice.

PART III

The Shepherd's Year

Chapter 8

The Late Summer – Weaning

The late summer is the start of the shepherd's year. The decisions taken at, and just after, weaning will lay the foundation for the lamb output and ewe prolificacy in the coming year.

When to Wean?

The decision must be taken on the basis of nutritional and parasitic considerations.

The nutritional background is that the ewe's lactation curve reaches a peak 3–4 weeks after lambing, and thereafter declines. By the time the lambs are 12 weeks old, the mother's milk is relatively unimportant and the lambs are relying on the grazing for most of their nutrients. But grass growth is on the decline from late June/early July and quality, too, can deteriorate at this time unless grassland management is of a very high standard, so the lamb becomes more dependent on grass at a time when it is in shorter supply. Moreover, there is increasing competition for the available herbage between ewes and lambs. Ultimately, the time comes when the ewe and lamb are in competition for the available grazing. The ewe's milk yield is declining and she eats the grass the lamb needs; clearly, the lamb will make better growth if weaned.

Except on clean grazing, worm infestation of pastures builds up in mid season. This can severely limit lamb growth, and the benefit from weaning and drenching lambs and turning them on to parasite-clean aftermath grass in early July is considerable.

It is sensible to delay the weaning of lambs which are close to finishing; otherwise, ADAS trials have shown no advantages in delaying weaning beyond 4 months. In a trial at Liscombe EHF, lambs weaned on to clean pasture at 12 weeks had higher 20-week weights than lambs weaned at 16 weeks (34.0 kg vs. 31.5 kg). The results have shown that better lamb-weight gains and swifter lamb sales have followed earlier weaning whenever the following conditions apply.

- When it is clear that there is excessive ewe vs. lamb competition for the available grazing.

- When parasite-clean aftermaths are available for the lambs.

Weaning earlier than 4 months can also benefit the ewe by allowing her more time to recover in body condition before mating. At the time of weaning, her body score may be around 2.5, whereas we know that a body score of 3.5 is desirable to optimise the number of eggs shed at mating time. It is therefore important to ensure that the ewe does not lose too much condition by late weaning, particularly in seasons of poor grass growth. If this occurs, the recovery in body condition may take as long as 12 weeks unless expensive compounds are fed.

The correct choice of weaning date will benefit both ewe and lamb performance. It will vary from season to season but the experimental evidence shows the following.

- Late weaning at over 16 weeks cannot be recommended except for lambs which are close to finish.
- Weaning at 12–16 weeks is normally preferable.

Management Post-Weaning

Lambs

At weaning, the lambs should be drenched against worms and moved to fresh, fertilised pasture. Hay or silage aftermaths may be available at this time of year, and are preferred for weaned lambs. The grazing should be chosen with care.

- It should be parasite-clean, either a new ley or grass not grazed by sheep during the previous 12 months.
- It should not be the grassland designated for grazing by ewes and lambs in the following season.
- The sward should be short (about 60 mm), leafy and clovery. Such grassland can be stocked at up to 50 weaned lambs/ha.

Ewes

After weaning, the ewes should be rapidly dried-off. This can be achieved on a sacrifice grass paddock (to be subsequently ploughed) where the ewes may be confined for 7 days at a density of 100–200/ha. Alternatively, the ewes may be housed and allowed only straw and water. This indoor method has two clear advantages.

- It is a speedy drying-off process which minimises the incidence of post-weaning mastitis.

- The ewes and lambs are out of earshot and both consequently settle down more quickly.

At weaning time, ewes vary widely in body condition. Some will have reared more lambs than others, and will therefore have milked more heavily. Prompt action is therefore needed to return each ewe to optimum body condition for mating.

The first step is to body condition score every ewe (see Appendix C) as soon as the drying-off process is complete. On the basis of these scores, the sheep should be divided into three flocks (for fat, fit and thin ewes, respectively) for differential treatment as follows.

Fat ewes (body score 4.0 or over)

The slimming down of over-fat ewes is controversial and, some say, unnecessary. However, there is evidence (see Chapter 9) that ewes going to the tup over-fit will have fewer lambs and that more will be barren. The slimming process may be attempted by dense stocking on grass or stubbles. However, this is surprisingly difficult to achieve and the authors prefer to keep the dried-off, fat ewes indoors for around 3 weeks on a straw-and-water regime.

It should be remembered that ewes over-fat at weaning are not usually amongst the most prolific or best-milking animals. Therefore, they are candidates for sale, particularly if the cull ewe price is attractive.

Fit ewes (body score 2.5–3.5)

Fit ewes require no special treatment in the period from weaning to flushing. Their body condition should be maintained.

Thin ewes (body score 2.0 or less)

This group is likely to contain most of the highly prolific ewes which have produced twins or triplets. They must be allowed the best grazing at a fairly light stocking rate to allow them to return to fit condition. If there is insufficient grazing available, then feed 0.2–0.5 kg/head/day of whole-grain cereal for 6 weeks prior to tupping and continue this through the tupping period.

Culling the Flock

Culling is best carried out at the time of weaning. Although prolificacy tends to decline after the fifth lambing, many ewes perform well until their seventh lambing and even beyond. Age alone is not therefore a

Table 8.1 Reasons for culling in a lowland flock

	(%)
Teeth	75
Udder	15
Other reasons	10

Source: Rosemaund EHF (1968–88).

good criterion for culling in the lowland flock. The following are the main reasons for discarding ewes (Table 8.1).

- *Teeth*. If the incisor teeth have been lost, special feeding will be needed and, if this is not possible, the ewe is normally culled. In the case of a young ewe losing condition but with good incisors, faulty molars should be suspected. They may be inspected by looking into the mouth, but this is difficult. It is easier to feel along the jaw line externally. If the molars form a smooth line all is well, but if there are irregular ridges this indicates thickening of the jaw bone. Such a ewe may be unable to chew her cud and should be culled.

  In lowland flocks some ewes culled because of poor teeth may be transferred to an early lambing flock and tupped again in August. Such ewes would then be sold after their lambs had been weaned.

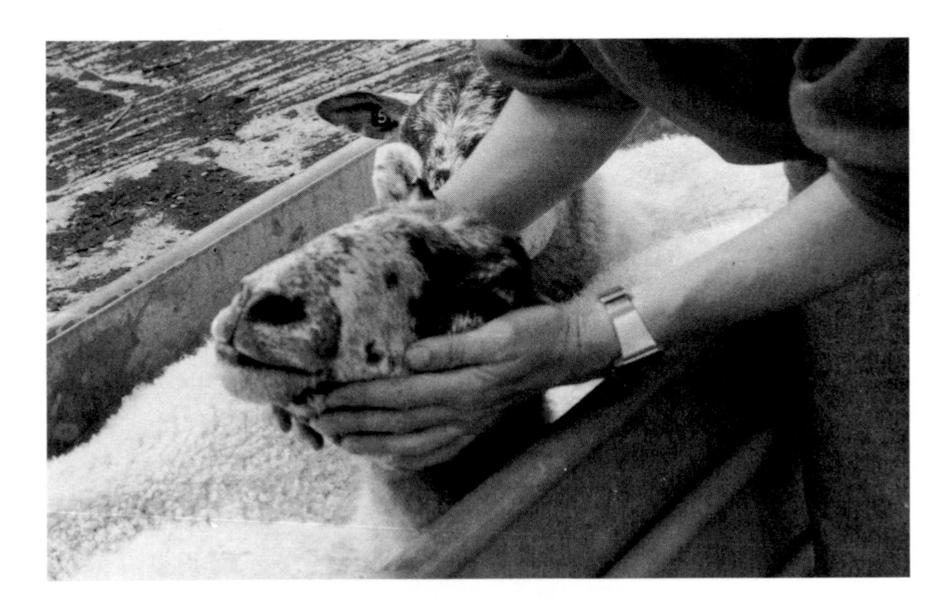

Plate 8.1 Feeling ewe's jaw for back teeth abnormalities

- *Udder*. All breeding ewes must have sound udders. Any with udder lumps or other abnormalities (probably already noted at lambing time) must be culled.
- *Feet*. Persistently lame ewes will eventually lose condition and are not worth keeping. Moreover, they may well be carriers of foot rot, and as such are a source of infection for the rest of the flock.
- *Condition*. Ewes in very poor condition because of disease and/or old age should be culled if it is unlikely that they can achieve a body score of 3.5 before tupping.
- *Breeding performance*. Older barren ewes should be culled in the spring as they tend to get over-fat during the summer and may be difficult to get in lamb again. Younger barreners are usually given a second chance.
- *Prolapse*. Experience shows that ewes with a vaginal prolapse will frequently prolapse again within 2 years. They should therefore be culled.

At Rosemaund EHF, the annual percentage of the ewe flock requiring replacement has varied over the years between 12 and 20, which means an average flock life of 6 years. The buying of replacement ewes is discussed in Chapter 2.

Chapter 9

The Autumn – Preparations for Tupping

INCREASING PROLIFICACY

The benefit of getting ewes into the correct condition for tupping (condition score 3.5) is increased prolificacy. This process should start as soon as the ewes have been dried-off, and should be completed by 2–4 weeks before mating starts. Then it is time to flush the ewes.

Flushing

Flushing is the practice of putting the flock on really good grazing from 2 to 3 weeks before tupping; this means the maximum number of eggs will be shed, thereby improving the lambing percentage, and this is supported by recent research carried out at the Macaulay Land Use Research Institute (MLURI). However, the practice of allowing ewes to lose condition after weaning so that they can subsequently benefit from a period of flushing is not recommended.

Although ewes already brought to the ideal body condition may not have any more lambs as a result of flushing, the technique has two possible advantages.

- Any ewes still a little on the lean side will perform better.
- Ewes fit at tupping stand a better chance of still being in good condition in late pregnancy, and this can mean heavier lamb birth-weights and less twin lamb disease.

The ewes should be kept on good grazing throughout the tupping period. Only if the grass is lacking in quantity or quality should concentrate feeding be considered at this time.

Ram preparation

Replacement tups, including one or two extras as 'reserves', should be purchased some 2 months before the commencement of tupping in order to allow a settling-in period.

Although the ram is half the flock (according to the popular saying) it is remarkable how little care is taken over ram selection. Some flockmasters consider that price is the main factor and make no physical examination of the ram before purchase. It is suggested that the flockmaster adopt the following check-list and apply it 8 weeks before tupping both to purchased tups and to those already in the flock.

- *Feet*. Following careful examination, any foot rot should be treated by paring and then dressing with the preparation recommended by the veterinary surgeon. Tups will not serve if their hind feet are sore.
- *Teeth*. Examine both incisors and molars for any abnormalities.
- *Conformation*. Check in particular the hind legs, feet and mouth.
- *Health*. Look for signs of infectious conditions such as orf, lice or conjunctivitis. All tups should be worm-drenched as a routine before tupping. Normally, adult sheep can cope with light worm burdens without showing adverse effects. However, during the tupping period, rams are under stress, spend little time grazing, and can lose condition rapidly if carrying a worm burden to end up looking like walking coat-hangers.
- *Fertility*. The testicles should be two in number (this is not always the case!) and firm with no lumps. The penis should be exposed and checked. Surveys have shown that 10 per cent of tups are infertile, so if there are any doubts a veterinary examination should be requested. In special cases pedigree breeders or those running a single tup with a group of ewes should consider asking the veterinary surgeon to do a semen test.
- *Condition*. At 4–6 weeks before tupping, the rams should be assessed for body condition. They should be body score 3.5 or 4 by tupping, and any thinner tups should be fed an 18 per cent crude protein (CP) concentrate at 0.5–1.0 kg/day.

Choice of ram breed

The authors believe firmly that the choice of breed for the terminal sire is not critically important. What is important is that whatever breed is chosen, the ram used should be a good example of that breed. There is just as much variation within a breed as between breeds. However, it is clear that the terminal sire breeds vary considerably in earliness of maturity, and therefore the choice must depend on the market requirement particularly in terms of lamb carcass weight. Many flockmasters consider the Suffolk to be an excellent dual-purpose sire, producing ideal lambs in the standard weight range, and also popular store lambs.

At Rosemaund EHF, the progeny of Hampshire, Suffolk and Dorset Down rams crossed on Mule ewes were compared. The surprising result of this observation was that up to weaning in July there was absolutely no difference in the performance of the lambs. After weaning there was a slight difference in that, whereas the Dorset Down cross lambs grew and put on fitness, the lambs by the Hampshire and Suffolk rams gained weight but became fit for sale more slowly.

Over the last few years, continental breeds have been imported into the UK for use as terminal sires. The Texel has proved to be one of the best sires for the production of heavy lambs of good conformation. Charollais cross lambs have also won many carcass classes in recent years. It must be remembered that, in the main, the better examples of these breeds were imported. Nevertheless, these importations of foreign breeds have done a lot of good if only to provide competition and prevent the pedigree breeders of traditional breeds in the UK resting on their laurels.

Vasectomised rams

Vasectomy is a simple operation which leaves the ram with its sex drive but renders it completely infertile. The operation is irreversible. This is an effective technique for tightening up the lambing period. It is less costly than the alternative synchronisation method of using hormone-impregnated sponges, and its use in early lambing flocks is fully discussed in Chapter 7.

Some flockmasters have reported improved synchronisation when entire rams have been run alongside the ewes (at the other side of a sound fence) prior to tupping. However, the teaser technique is more effective and there is less risk of accidents!

A final point. The technique concentrates the tupping period. So it is important to run not more than 30 ewes/tup when mating commences.

Raddling the ram

If a colour dye mixed with oil is applied to the ram's brisket, the ewe is marked on the hindquarters after being mated. The colour of the raddle should be changed at least every 17 days. However, many flockmasters now prefer to change the colour every week. The advantage is that ewes which are subsequently housed can be penned according to the expected week of lambing as indicated by the raddle colour. More accurate feeding of pregnant ewes is thus made possible.

Moreover, the changes in raddle colour clearly indicate ewes which return to the tup. If there are too many of these, the fertility of the rams should be investigated. The final colour change, usually to black, is normally made after 36 days, and in the interests of a concentrated lambing period, ewes marked black should be culled.

An alternative is to fit the rams with a harness into which a coloured crayon is slotted. This has the advantage that the rams do not have to be frequently caught to be re-raddled since a colour crayon will last throughout a 17-day period. The disadvantage is that the harness gradually beds into the wool and may also become loose. It must therefore be checked regularly to make sure that it is not chafing the skin. If this happens it can cause the ram serious discomfort and abscesses may develop, both of which will prevent him from serving ewes.

Tupping management

The objective here is to maximise the number of lambs born and minimise the number of barren ewes. With this in mind it is false economy to try and make do with too few tups. There should be 1 mature ram/40 ewes and 1 ram lamb/20 ewes.

The size of the tupping group is important. Small groups of ewes depending on one tup may be necessary in very small flocks but are far from ideal (since many barreners can result if one ram's fertility is low) unless rams are rotated between groups. Likewise, groups containing two tups are not to be recommended since, in the authors' experience, this frequently results in fights for supremacy and the consequent occasional loss of a ram with a broken neck.

At Rosemaund EHF, the best results have come from running three tups with a group of a hundred ewes. Fighting has then been rare. Larger tupping groups of around 200 ewes have reportedly given poorer results.

Ewe lambs should be run with mature tups and ram lambs with mature ewes, the theory being that it is unwise to put inexperienced individuals together. In practice, a well-grown ram lamb may be a successful member of a trio of tups running with around a hundred ewes.

If tupping is carried out on good-quality grassland, supplementary feeding of the tups is normally unnecessary. However, if grass keep is a bit short the tups should be fed a 16–18 per cent CP compound throughout the mating period. Some flockmasters train their rams to take feed from a bucket, and this makes such feeding easy.

Artificial Insemination (AI)

The use of artificial insemination (AI) is not popular with UK flockmasters, although it has been estimated that over 60 million sheep are inseminated annually worldwide, mainly in the USSR and Eastern European countries.

Semen collection

The usual method of semen collection is to train the ram to mount a teaser ewe and ejaculate into an artificial vagina; this technique can be carried out by anyone possessing a degree of patience. The second method, electro-ejaculation, is performed only by skilled operators and animal welfare considerations are paramount; this method is useful for checking the semen quality of a ram before the mating season commences.

Collecting into an artificial vagina provides high-quality ejaculates, especially where a number of collections are required over a short period. Immediately after collection, the semen should be examined under a microscope to check sperm motility. If the semen is to be used within 20 minutes it can be divided into 0.05–0.1 ml doses and used undiluted. If the semen is to be stored, it is diluted 1 part semen : 2 parts dilutant containing skim milk and antibiotics. It is then cooled from blood temperature to 15 °C over 30 minutes and can be stored at this temperature for up to 10 hours before use.

Dose rates of 125–150 million sperms are required for ewes in natural oestrus (ewes detected by teaser rams) but if ewes have been given a progesterone sponge, dose rates of 400 million sperms are required. The reason for this is that progesterone is slightly spermicidal and even 2 days after sponge removal the progesterone has an adverse effect on sperm motility.

Rams as well as ewes are seasonal breeders and the time of year has a marked effect on the number of sperms per ejaculate. During trials conducted by the MLC with adult Suffolk rams it was shown that outside the normal breeding season approximately 5 collections can be taken per week but in the breeding season this increases to 20 collections per week.

On average, rams will produce between 3000 and 4000 million spermatozoa per ejaculate and this will provide between 7 and 20 doses of semen, depending on the rate used.

Timing

In the UK most of the ewes to be inseminated are synchronised by the use of progesterone sponges and injected with PMSG. The intra-vaginal sponge is removed after 12–14 days and the PMSG is given at the time of removal. Normally, the ewes come into heat 2 days later. If a single insemination is to be made, the best time is 56 hours after sponge removal. Where a double insemination is to be made the timing is not so critical, and the inseminations are normally timed at 50 hours and 60–64 hours after sponge removal.

For cervical inseminations, the ewe is usually restrained in a standing position with the hind-quarters slightly raised. An illuminated duck-billed speculum is inserted into the vagina and a pipette passed along it so that the end of the pipette locks into the cervix. The spermatozoa are released on to the surface of the cervix. The cervix provides a major barrier to the passage of sperm into the uterus and in the ewe it is not possible to pass the end of the pipette through the cervix to discharge the sperms inside the uterus. Hence, large numbers of sperms have to be released on to the posterior wall of the cervix or into the vagina. However, skilled operators can achieve results equal to those resulting from natural service.

Surgical methods of placing the sperm directly into the uterus have not proved very successful, but recently a laparoscopy technique, developed in Australia and in the UK, is proving to be more promising. This method involves inserting a fibre-optic laparoscope via a canula in the abdominal wall so that the uterus can be viewed. Semen is carefully injected through a second canula in the abdominal wall, into each of the two uterine horns. This technique has the advantage that only 10 per cent of the sperm required for cervical insemination are necessary, but it can be performed only by those with a Home Office licence.

Deep-frozen semen

Unfortunately, ram's semen at the present time does not freeze as successfully as bull's. Experiments are continuing to find ways of improving the freezing techniques, and conception rates of over 50 per cent have been achieved using semen which has been deep-frozen.

Advantages of using AI

- There is a reduction in the number of rams required, especially advantageous for a synchronised flock.

- Pedigree breeders can use valuable proven rams more extensively.
- The carcass quality of finished lambs can be improved by the widespread use of rams with good conformation and/or with high growth rates and good lean meat : fat ratios.
- Artificial insemination reduces the risk of buying in disease since fewer new rams are purchased. In this connection it is interesting that the Swaledale Sheep Association is co-operating with ADAS and the MLC in a long-term project with the objective of breeding scrapie-resistant sheep, initially in seven flocks. Ewes in these flocks are inseminated using semen from older rams known to be scrapie-resistant.

Breeding from Ewe Lambs

Experimental work conducted with ewe lambs has been limited, mainly because they seem to be a law unto themselves. In some years at Rosemaund EHF, results have been good, with lambing percentages over 120. In other years, when both the breed types and management have been unchanged, the lambing percentage has been between 50 and 60.

There is a financial advantage in breeding from ewe lambs. They can be purchased for about £20–30 less than replacement shearlings, and during their lifetime on the farm they should produce more lambs (Table 9.1). The lambs produced by ewe lambs are regarded

Table 9.1 Average number of lambs produced during lifetime

	Ewe lamb	*Shearling*
Bred as ewe lamb (first lambing)	1	—
Bred as shearling (second lambing)	2	2
Third lambing	2	2
Fourth lambing	2	2
Fifth lambing	1	1
Total	8	7
Purchase price	£60	£80
Ewe replacement cost/lamb produced	£7.5	£11.4

Source: Rosemaund EHF.

as a bonus since with good management, their subsequent breeding performance is similar to that of ewes first bred as shearlings.

The additional lamb produced by the bred ewe lamb reduces the flock overheads in terms of ewe replacement cost. It is also widely accepted that a ewe which has bred as a ewe lamb makes a better mother when bred as a shearling, and for this reason bred shearlings are normally worth more than unbred ones.

It is known that both age and weight of the ewe lambs at mating have a great influence on whether or not they will come on heat and conceive.

In self-contained flocks, it is better to retain the earlier born ewe lambs as flock replacements. If ewe lambs are purchased for breeding in their first year it is usually wise to pay the extra money and buy well-grown animals. It is recommended that ewe lambs should have reached 65 per cent of their mature weight by mating time.

Ewe lambs should be purchased at the early sales so that they have time to settle down on their new farm. It is often forgotten that these purchased ewe lambs may have been reared in hill or upland areas and that the sudden change to lush grass/clover pastures can be stressful. On arrival, purchased ewe lambs should be drenched against worms and fluke and receive an injection against clostridial diseases and pasteurellosis. They are then best turned out to bare pastures and left to settle down.

In late September they can be put on to better pastures. A mineralised cereal should be fed if grass is scarce before or during tupping. The tups are normally turned into the ewe lambs (1 : 30) 3 weeks after they are turned in with the ewes. This allows the ewe lambs an extra 3 weeks to grow and mature and it also means that they commence lambing at a time when the lambing of the mature ewe flock has passed its peak.

Chapter 10

The Winter – Pregnancy

Managing the Pregnant Ewe

Month 1

Ewes in spring-lambing flocks are normally out at grass in early pregnancy. If rested swards have been made available and the stocking rate is limited to 6–8 ewes/ha, the flock should have forage to appetite of a reasonable nutritive value (say 60 'D') and no supplementary feeding will be required. Such forage should be freely available until early December to prevent embryo loss.

However, where tupping is delayed until November, a supplement of around 500 g cereals/ewe/day may be needed. This will also apply where forage is short or stocking rate is high. Ewe lambs will require 200–300 g cereals/head/day supplement to allow for continuing maternal growth. Supplementary feeding of the whole flock is prudent during prolonged spells of cold, wet, November weather.

To summarise, there are two golden rules covering the management of ewes during the first month of pregnancy, namely:

- Keep the ewes on their pre-tupping plane of nutrition.
- Avoid all flock disturbance and changes in environment.

Months 2 and 3

During November and December, ewes in mid pregnancy can graze around the available grass fields eating the last of the autumn grass, and may well not require supplementary feed. However, from the turn of the year, additional food (usually in the form of hay in racks) is likely to be necessary, particularly in bad weather. Since total feed intakes at grass are almost impossible to estimate, this additional feeding is a very useful insurance to avoid any danger of under-feeding that exists at this time. This risk is reduced if the ewes are housed and all feeding is under the control of the flockmaster. It is advisable to feed hay or silage at grass for a week or so before housing the ewes in order to make the change in feed regime more gradual. This is particularly important for young ewes or ewes which have not been housed in previous winters.

During mid pregnancy, over-feeding is probably a greater danger than under-feeding. For ewes in body condition 3 or over, winter grazing (supplemented from January if necessary to maintain body condition) or its equivalent of hay or silage if indoors, is adequate. Extra feeding of such ewes is wasteful, and can also be dangerous. Over-feeding leads to over-fat ewes in late pregnancy which in turn predisposes them to twin lamb disease and other nutritional disorders and to prolapse and lambing difficulties.

Ewes in poor condition (body score 2.5 or less) and ewe lambs should be grazed separately at a lower stocking rate and allowed up to 1.5 kg hay/ewe/day in racks. If housed and fed straw or moderate quality hay or silage they should be allowed a daily supplement of 500 g of a cereal-based compound containing 15 per cent soya bean meal or 10 per cent fish meal with an appropriate mineral/vitamin supplement.

The key management points in mid pregnancy are as follows.

- Condition-score the ewes regularly.
- Do not allow a loss in condition of more than half a unit in this period.
- Feed ewes in satisfactory condition (body score 3–3.5) on a maintenance diet to avoid over-fatness.
- Allow extra feed to ewes in poor condition, to ewe lambs and to the whole flock if the winter grazing is inadequate or conserved forages available are of below-average quality.
- Consider pregnancy scanning. The pros and cons are detailed later in this chapter.

Months 4 and 5

If the flock is out-wintered there is always a danger of overestimating the feed value of the grazing. Supplementary roughage should be offered, and a cereal/protein mix will be required for the last 8 weeks before lambing.

Housed ewes can be more accurately fed, particularly if the opportunity is taken to pen them according to their feed requirements. Penning arrangements are fully discussed later in the chapter.

In order to feed the heavily pregnant ewe according to her requirement but as economically as possible, it is essential that the forages to be used are analysed beforehand. With most hays and silages it

Table 10.1 Rations for the 70 kg lowland ewe in last 8 weeks before lambing

	6–8 weeks (kg/day)	*4–6 weeks* (kg/day)	*2–4 weeks* (kg/day)	*0–2weeks* (kg/day)	*Total* (kg)
High-quality hay	1.40	1.60	1.50	1.25	80.5
Compound	—	—	0.25	0.70	13.5
Good silage	4.00	4.50	4.50	4.50	245.0
Cereals	—	—	0.15	0.35	7.0
Straw	1.50	1.20	1.00	0.80	63.0
Compound	0.55	0.75	0.91	1.10	46.0

NB: The compound fed with hay should contain 14–16 per cent crude protein (CP). The compound fed with straw should contain 16 per cent CP and be fed from housing.
Source: Based on ADAS Leaflet 636, *Feeding the In-lamb Lowland Ewe*.

is necessary to feed a supplement to the ewes in the last 4–8 weeks before lambing. The objective is to increase the energy concentration of the ration when the ewe's requirement is rising rapidly and rumen volume may be reducing because of rapid foetal growth.

Most flockmasters increase the ration of cereal-based concentrate weekly in 'steps' in line with requirement. The translation of the theoretical requirements of ewes carrying twins detailed in Chapter 4 gives the practical rations shown in Table 10.1.

Where the hay or silage on offer is of very high quality, a supplement of whole cereals with a mineral/vitamin addition may be adequate. However, with most forages a purchased concentrate (14–16 per cent CP) or a home mix with a similar protein content is required (see Chapter 4). There is now considerable evidence that if, in late pregnancy, the gap between the energy requirement and the energy supplied in the diet widens, it is advantageous to feed some high-quality protein. Many flockmasters include fish meal or soya bean meal at 5 and 7.5 per cent, respectively, of the concentrate by weight from 6 weeks before lambing, increasing this to 10 and 15 per cent 2 weeks before lambing. The minerals and vitamins necessary will be contained in purchased compounds, but flockmasters preparing home mixes should never forget that ewes heavy in lamb have critical requirements for minerals, trace elements and vitamins. A mineral/vitamin supplement formulated for sheep should be mixed with the other constituents at the rate recommended by the supplier. However, this inclusion rate should not be exceeded, nor should free access to minerals be allowed, as the delicate balance can be upset, leading to hypocalcaemia.

Appetite in late pregnancy

All possible measures must be taken to encourage the flagging appetite of the heavily pregnant ewe, including the following.

- Provide plenty of trough frontage for the feeding of concentrates, namely 450–500 mm/ewe.
- Where silage is fed, renew the supply twice daily and remove all uneaten feed from the troughs daily. *Never* change silage clamps in late pregnancy, even if the change is to a better-quality forage. Appetite can be adversely affected, resulting in cases of twin lamb disease.
- Limit the weight of compound fed to 500 g/ewe/feed. This could mean twice- or even thrice-daily feeding. Make any changes in ration constituents gradually.
- Have clean water available at all times.
- Winter-shear the ewes where appropriate (see later in this chapter).

Some authorities recommend that forage should be rationed in late pregnancy in order to encourage the intake of concentrates and reduce the risk of prolapse. The authors disagree, since the result is that the ewes get a more starchy and less fibrous diet. The two most likely outcomes of this policy are digestive problems and large bills from the feed compounders.

The important points in the management of the heavily pregnant ewe are as follows.

- Avoid abrupt changes in management or feeding. If ewes are to be housed, this is best done at least 8 weeks before lambing.
- Maintain body condition with no more than a half unit loss.
- If possible, pen the ewes according to their feed requirement.
- Reserve the best-quality forage for this period and feed it to appetite.
- Base supplementary feeding on forage analysis and ewe condition. In planning feed levels, assume all lowland ewes are carrying twins.

Aids to Winter Management

A number of rewarding techniques have been developed in recent years. All can promote the flockmaster's objectives of producing more lambs of satisfactory birth-weight, reducing ewe and lamb mortality and achieving economies in feed use. The most important are body condition scoring, separate penning, pregnancy scanning and winter shearing. A further important aid to accurate feeding is the checking of feed intakes of sample pens of ewes every 2 weeks.

Body condition scoring

Body condition scoring is described in Appendix C. Body scoring gives perhaps the best indication of the ewe's health and well-being. It is normally more meaningful than live-weight, e.g. a pregnant ewe carrying twins may be 15 kg heavier just before lambing than she weighed at mating. But the extra weight is fully accounted for by the foetuses and associated membranes and fluids. The body condition score could have declined by 0.5–1 unit in the same period, and is therefore a more accurate indicator of the adequacy of the feeding regime than body weight.

Condition scoring should be part of the normal management of the flock. Body score targets should be met throughout the year by appropriate management.

Separate penning

Sheep housing allows pregnant ewes to be separately penned in groups of 30–40. Ewe lambs and shearlings are less likely to be bullied if separated from their more aggressive elders. Ewes may also be penned on the basis of the following considerations.

- *Body condition*. But steps should have been taken to even this up during the autumn, and body condition should therefore be fairly uniform by the time the ewes are housed.
- *Number of lambs expected*. This applies if the ewes have been pregnancy-scanned, when ewes carrying singles can be separated from ewes carrying multiples. In the absence of scanning, single-bearing ewes may be over-fed, and those carrying triplets under-fed.
- *Expected lambing date*. This is the most rewarding basis for separate penning in flocks with first-class management. It is made possible by the raddling of the rams at tupping time. At Rosemaund EHF, the colour of the raddle crayon is changed weekly and the ewes are penned at housing according to the colour of their marking. The result is that in each pen, most ewes will lamb within a period of 14 days. This may mean extra supervision to prevent mis-mothering, but the accuracy of feeding is greatly enhanced. Earlier-lambing ewes are less likely to be under-fed and later-lambing ewes are less likely to be over-fed. This in turn means more economi-

cal feed usage, more even lamb birth-weights and lower lamb mortality.

It is also important to have a hospital pen, especially for any ewes with chronic foot rot.

Pregnancy scanning

Until recently, the available equipment could diagnose pregnancy but was inaccurate in 'counting' the number of foetuses. However, the real-time ultrasonic scanner can accurately identify foetal numbers. This scanner is similar to the equipment used in maternity hospitals and pre-natal clinics. The ewe is held on a cradle and, following a light application of vegetable oil to the belly, the probe of the scanner is moved over the skin immediately in front of the udder. A beam of high-frequency sound is passed through the ewe and the echoes reflected back to the probe are converted to show images of the abdominal contents on a small television screen. An experienced operator can count the number of foetuses present in about 1 min/ewe, so 500 ewes may be scanned in a working day. Scanning is completely harmless to the ewe, the unborn lambs and the operator.

Apart from securing the services of a reliable and experienced operator, the flockmaster must make the following preparations for scanning.

- Present the ewes for scanning 50–100 days after the commencement of tupping.
- Starve overnight.
- Prepare a dry covered area for scanning, with a 13 A electric point within reach.
- Be prepared to identify each ewe by a colour mark as a barrener or as carrying a single lamb, twins or triplets.

In a 1983 trial by the MLURI in conjunction with the MLC, the correct foetal number was identified in all but four of 285 Scotch Halfbred ewes with a 200 per cent lamb crop.

A 1985 ADAS survey in Powys of 102 flocks with almost 38,000 ewes revealed an overall accuracy of 94.2 per cent.

Recent reports indicate that barren ewes should be identified with almost 100 per cent accuracy. The number of lambs carried should be predicted with 90–95 per cent accuracy.

Advantages of scanning

A main advantage is undoubtedly the early identification of ewes which are barren. Such ewes can then be sold at the relatively high price obtainable in late winter and before they consume expensive concentrates. A second advantage on the hills is the ability to bring twin-carrying ewes into the in-bye earlier than those carrying singles.

The technique also makes possible a saving in feed costs arising from the separation of those ewes carrying singles from those carrying multiples – this saving should be £2–3 per single-bearing ewe. Where housing is limited, it is advantageous to be able to identify and house the ewes carrying multiples.

Pregnancy scanning can improve the accuracy of feeding in late pregnancy. Ewes carrying singles can be fed less, thus avoiding excessive weights of single lambs, and this means fewer difficult lambings; this is particularly important with ewe lambs and in early lambing flocks, two situations where the lambing percentage may be relatively low. Similarly, ewes carrying multiples can be more generously fed, resulting in a lower incidence of twin lamb disease, a higher average lamb birth-weight and therefore lower lamb mortality.

It should be noted, however, that these advantages will only be achieved if the means exist (either at grass or indoors) to separate ewes into groups on the basis of the scanning results.

Disadvantages of scanning

The disadvantages of scanning are the cost of the scanning contractor, extra labour and the extra handling of ewes. Some flockmasters claim that the money spent on scanning would be better spent on buying extra concentrates for the whole flock. Most estimates of the financial advantage of the technique have shown this to vary from nil to £4/ewe. A careful costing carried out by Redesdale EHF showed a benefit to scanning of £3.18/ewe.

In summary, it seems that the technique has least to offer to flockmasters with highly prolific flocks where the lambing percentage approaches 200. This is because such flocks will have few ewes barren or carrying singles, and therefore the information gained from scanning may be of little value. However, for flocks lambing at 120–170 per cent, pregnancy scanning is a new and exciting aid to the management of the pregnant ewe. Many flockmasters have incorporated it as a part of their routine flock management.

Winter shearing

Early development work on the January shearing of housed pregnant ewes was carried out at Great House EHF in 1971 and 1972 and was regarded by the authors (and most others) as a gimmick of little significance. Subsequent trials at Drayton, Liscombe and Rosemaund EHFs and in many commercial flocks have proved that the handling and clipping of ewes in mid January gives few problems. Moreover, the technique confers three major advantages: higher birth-weights, a lower requirement for house space and greater ewe comfort.

Higher birth-weights

It should be noted that winter shearing does not result in extra lambs being born. Table 10.2 shows results from three commercial flocks and from Rosemaund EHF quoted by H. Morgan and J. S. Broadbent. However, the beneficial effect of winter shearing on lamb birth-weight is well illustrated in Table 10.3.

It is well known that there is a strong positive correlation between lamb birth-weight and lamb survival (Figure 10.1). The 0.5 kg advantage in birth-weight secured by winter shearing (found also at Drayton EHF) therefore means lower lamb losses.

Table 10.2 Litter size and live births per ewe

	Winter-sheared ewes		*Ewes not winter-sheared*	
	Litter size	*Live lambs per ewe*	*Litter size*	*Live lambs per ewe*
Farm 1	2.02	2.00	2.19	2.10
Farm 2	2.03	2.00	2.08	1.96
Farm 3	1.87	1.83	1.90	1.87
Rosemaund EHF	1.86	1.68	1.90	1.90

Source: British Society of Animal Production, winter meeting 1980.

Table 10.3 Birth-weights of lambs 1979–81

	1979		*1980*		*1981*	
	Shorn (kg)	*Unshorn* (kg)	*Shorn* (kg)	*Unshorn* (kg)	*Shorn* (kg)	*Unshorn* (kg)
Singles	6.1	6.5	6.2	5.2	6.6	5.0
Twins	5.2	4.5	4.5	4.2	5.0	4.4
Triplets	4.6	4.0	3.5	3.2	3.7	3.8
All lambs	5.3	4.6	4.5	4.1	4.9	4.3
Percentage difference	+8		+12		+14	

Source: Rosemaund EHF.

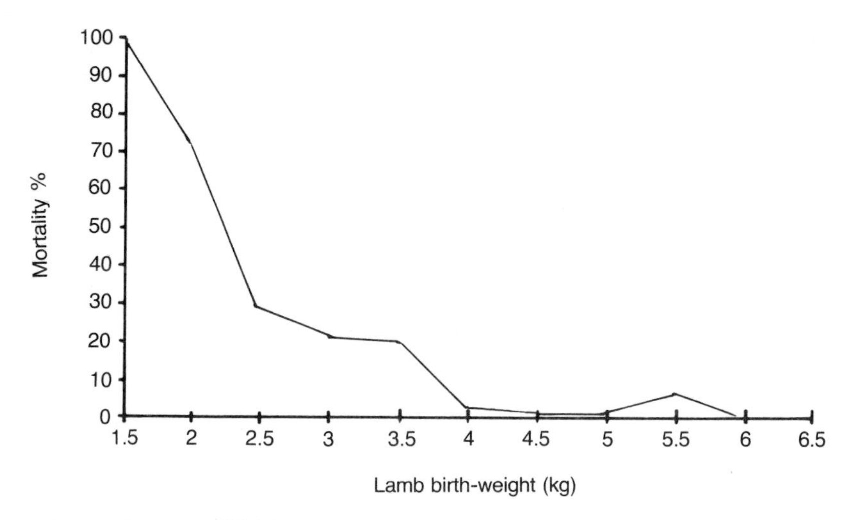

Source: Rosemaund EHF

Figure 10.1 Relationship between lamb birth-weight and mortality to 28 days

In the highly prolific flock with many multiple births and a relatively low average lamb birth-weight, the improvement in lamb viability conferred by January clipping is particularly important. Indeed, many flockmasters have commented that the principal advantage of January shearing is that it results in more evenly sized twins and triplets.

Better birth-weights also mean bigger lambs at weaning. At Rosemaund EHF, lambs out of winter-shorn ewes have had a 3–5 per cent weight advantage at 28 days of age.

A lower requirement for house space

At Liscombe EHF shorn ewes housed at 0.74 m^2 floor space/ewe performed as well as similar unshorn ewes stocked at 1.02 m^2/ewe and the consensus of the evidence is that stocking density in the house may be safely increased by 20 per cent following January clipping. This is an important advantage because it reduces the housing cost of the flock. It has also been claimed that trough space for compound feeding may be reduced by 50–100 mm/ewe. However, this is a more critical allowance than floor space because of its effect on feed intake, and economies in trough space should always be viewed with extreme caution.

Greater ewe comfort

The increased rate of metabolism in the ewe in late pregnancy means the production of more heat, and when insufficient heat is dispersed

Table 10.4 Respiration rate/minute at housing and 1 month pre-lambing

	At housing			*1 month pre-lambing*		
	House temperature (°C)	*Shorn*	*Unshorn*	*House temperature* (°C)	*Shorn*	*Unshorn*
Farm 1	−1.5	25	53	+3.5	29	73
Farm 2	+2.0	25	60	+7.5	36	172
Rosemaund EHF	+2.0	27	54	−0.5	33	65

Source: British Society of Animal Production, winter meeting 1980.

through the fleece both skin temperature and the respiration rate rise. The rise in the respiration rate of unshorn ewes can be quite dramatic in late pregnancy during periods of mild weather. However, the January shearing of housed ewes has been shown to halve the respiration rate. The figures in Table 10.4 were given by Morgan and Broadbent.

In addition to higher birth-weights, lower requirements for house space and greater ewe comfort, there are further points in favour of January shearing, namely:

- It is carried out at a quiet time of the farm year, whereas summer clipping may well clash with the silage- or haymaking seasons.
- The body condition of the shorn ewe can be better judged, allowing appropriate changes to be made in the feeding regime.
- Lambing problems are more easily recognised, and sucking is easier for the new-born lamb because it can more readily find the teat.

Winter shearing has two main disadvantages: shorn ewes eat more and there may be more lambing difficulties.

Shorn ewes eat more

There is no doubt that winter-shorn ewes have a higher feed intake, perhaps in part to keep themselves warm. The forage dry matter (DM) intakes in Table 10.5 were recorded at Rosemaund EHF and reported by Morgan and Broadbent.

Results at Rosemaund EHF with ewes fed silage showed that winter-shorn ewes ate 12–16 per cent more forage DM than unshorn ewes. At Drayton EHF where the ewes were fed hay, shorn ewes showed a 12 per cent increase in feed DM intake.

January shearing increased feed costs at Rosemaund EHF by around £2/ewe. However, Table 10.5 shows that this higher feed intake

Table 10.5 Actual and theoretical feed intakes

Weeks after housing	*Shorn*		*Unshorn*		*Theoretical requirement of 70 kg ewe with twins*	
	Silage DM/head/day (kg)	ME	*Silage DM/head/day* (kg)	ME	*DM/head/day* (kg)	ME
1	1.51	16.2	2.13	22.8	0.51	10.2
2	1.27	13.6	0.96	10.3	1.03	11.0
3	1.75	18.8	1.29	13.8	1.11	11.9
4	1.48	15.8	0.87	9.3	1.20	12.8
5	1.99	21.3	1.67	17.9	1.28	13.7
6	1.29	16.7	1.21	15.8	1.38	14.8

Source: British Society of Animal Production, winter meeting 1980.

ensured adequate nutrition, whereas unshorn ewes were occasionally below the theoretical requirement, which meant some danger of twin lamb disease.

There may be more lambing difficulties

Ewe lambs and flocks with average or below-average prolificacy will have many single lambs. These will have an above-average birth-weight, and any further increase is unwelcome. At Rosemaund EHF, winter shearing has resulted in a greater number of difficult lambings of single lambs.

Two additional minor disadvantages of the technique should be mentioned.

- A reduced wool price for the first shearing following only 7 or 8 months of wool growth. At subsequent annual January clippings, wool quality has been satisfactory, providing there is no straw contamination.
- Shorn ewes are more difficult to catch, and vaginal prolapses are less easy to treat.

The management of winter-shorn ewes requires careful planning to avoid the possibility of cold stress immediately after shearing and before or after turn out, which can result in wool slip.

Shearing should be carried out at least 8 weeks before turn-out to grass to ensure that 12 mm of fleece has grown by that date. A standard comb leaves 4–5 mm of fleece and has been found to be satisfactory.

Draught-free housing is essential, and the availability of dry, clean bedding must not be skimped.

Shorn ewes must be turned out to grass in favourable weather, and fields with natural shelter, such as woods on two sides, should be chosen where possible. If such fields are not available, simple shelters should be provided in exposed situations.

To summarise, winter shearing should not be contemplated on the hills or for early lambing flocks, and only fit ewes (body score 2.5 or higher) should be shorn. The technique will promote better lamb birth-weights and survival and lower housing costs because of the higher stocking density indoors of shorn ewes.

In spring lambing lowland flocks with average or above-average prolificacy, winter shearing of housed ewes is an acceptable husbandry practice which can be recommended confidently.

Blood testing

The levels of certain metabolites in the blood give a good indication of the energy status of the ewe and thereby of the adequacy of the feeding. Flockmasters who are worried about the nutritional status of their ewes in late pregnancy could request their veterinary surgeon to take blood samples for analysis.

Health During Pregnancy

Doubts have frequently been expressed concerning the health hazards to which housed pregnant ewes are supposedly exposed. After some 30 years of keeping ewes indoors in the winter, the authors see no reason why such ewes should be any less healthy than out-wintered sheep. Of course, due attention must be paid to house ventilation and good hygiene, and there must be adequate provisions of feed and dry bedding. In addition, precautions should be taken against the following health problems experienced during the winter months.

Foot rot

The foot rot causative organism survives only in diseased feet and cannot live for more than 2 weeks in the soil. It is therefore theoretically possible to eliminate foot rot from a closed flock after weaning time; this means an intensive programme of five or six examinations of the feet of all ewes at weekly intervals and treatment of all diseased feet. Whether or not complete eradication of the disease is attempted, all sheep should have their feet trimmed and if necessary treated against foot rot before housing. They should be housed with sound feet, and subsequently foot-bathed at 14-day intervals in a 5 per cent formalin solution or in a zinc sulphate solution prepared in accordance with the manufacturer's recommendations. They should then be returned to freshly bedded pens.

Sheep with any foot rot at housing should be penned separately and treated until cured. Any with incurable foot rot should be culled. Foot rot vaccine is available as an injection. It will help to control the disease, and will reduce the substantial labour requirement of regular foot treatments.

External parasites

Lice can be a problem in housed sheep although they should have been eliminated by the compulsory autumn sheep dip. If they appear, a louse dust should be applied.

Stomach and intestinal worms

It has been common practice to dose ewes with an anthelmintic at lambing time to control the post-partum rise in worm egg output. However, several research workers have reported that the earlier drenching (pre-lambing) of ewes resulted in an increased lambing percentage. It was therefore decided to test at Rosemaund EHF the effects of dosing housed ewes in mid pregnancy 10 days after they were housed in January. Such a dose would remove any worm burden existing at housing, and since it seems unlikely that the ewes would be re-infected during their housed period, the winter drench could replace the drench formerly given at lambing time. Moreover, since winter feeding is expensive we took the view that this feed should be solely for the benefit of the ewe, and not for the benefit of parasitic worms.

Four experiments were conducted at Rosemaund EHF in the winters of the period 1978–81 involving 450 Welsh Halfbred and Mule ewes each winter. The ewes were divided at random into two groups. The ewes in one group were drenched with Nilverm (Levamisole) at 7.5 mg/kg and those in the other group were left untreated. In all other ways the two groups were managed identically.

The winter drench improved the lambing percentage in all 4 years as shown in Figure 10.2.

The overall averages were 193 per cent after drenching and 183 per cent for those not dosed. There were no significant effects on lamb birth-weights or on weight or condition of the ewes.

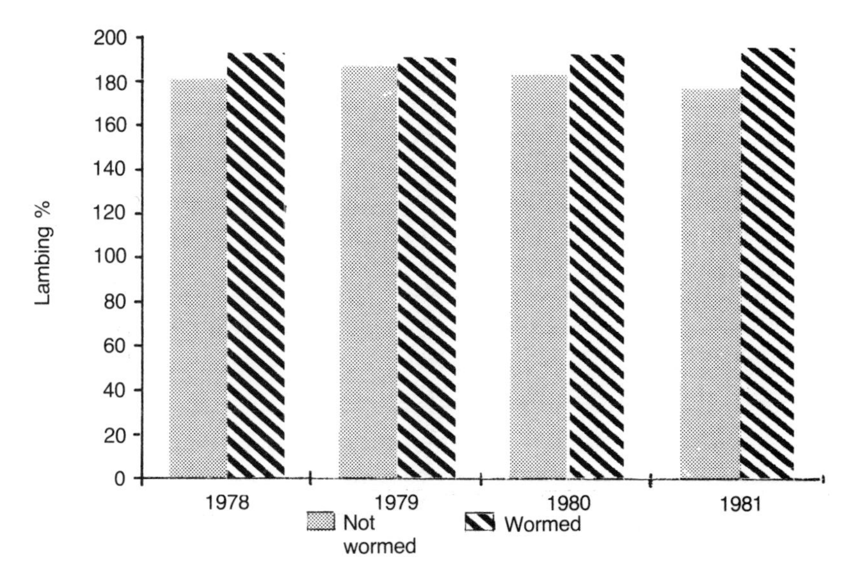

Source: Rosemaund EHF

Figure 10.2 The effect of worming ewes at housing on lambing percentage

The results at Rosemaund EHF were consistent but difficult to explain. It is possible that worm infestation in the untreated ewes adversely affected feed intake or utilisation which in turn increased the stress on the ewes and therefore the incidence of foetal deaths.

Although subsequent trials on commercial farms failed to give consistent results, the winter drenching of housed ewes is now widely practised. Since ewes are normally dosed once each year the changing of the date of this involves no additional expense. Moreover, the technique reduces by one the number of operations to be carried out during the busy lambing time.

Coccidiosis

Coccidiosis can be a particular problem in housed sheep, since infection may be picked up from the bedding. The control measures are fully outlined in Chapter 12.

Pregnancy toxaemia (twin lamb disease)

Twin lamb disease is caused by a failure to satisfy the ewe's energy demands in late pregnancy. This results in a low blood sugar level and an increase in blood ketones as the ewe mobilises energy by breaking down body fat. The nervous system becomes affected and coma and death may follow. The disease occurs most frequently in either over-fat or very thin ewes. Feeding of a high level of compound in a single feed can cause acidosis which in turn reduces appetite and leads to twin lamb disease. However, most cases are in ewes which are losing body condition, and the feeding in late pregnancy of low-quality roughages with little or no compound supplementation is a recipe for disaster.

The aim must be to prevent twin lamb disease by careful management pre-tupping and throughout pregnancy. Prevention is easier to achieve when the ewes are indoors, as the flockmaster has more control of the diet. This is why the disease is relatively rare in housed ewes. Important points in prevention are as follows.

- Feed ewes in the final 8 weeks of pregnancy so that they maintain weight. The target condition score is 3.5.
- Allow plenty of trough space and feed whole-grain cereals with a limit of 500 g compound/feed.
- Avoid changes in management and sudden variations in feed intake. This is particularly important in prolific flocks.

Chapter 11

The Spring – Lambing

Spring is a time of year when stockmanship is of paramount importance. Recent improvements in breeding and husbandry have made higher lambing percentages possible. This in turn means more multiple births and therefore a higher proportion of small lambs, so meticulous preparations for lambing and a routine planned in great detail are now more important than ever before. Otherwise, our successes in breeding more lambs per ewe will be thrown away because of unacceptably high lamb mortality at, and soon after, birth. In particular, the extra labour requirement at lambing time must be accepted if lamb mortality is to be kept low. Students from veterinary colleges are ideal if lambing occurs during their spring vacation.

Lamb Losses

A survey carried out by the University of Exeter from 1971 to 1974 indicated that the annual loss of lambs in the UK could be around 2.8 million, and ADAS surveys have concluded that lamb mortality may be approximately 3 million/year and that one lamb in five dies within a week of birth. These figures are a serious cause for concern because the higher the lamb mortality the fewer the lambs reared and sold and the lower will be the flock output.

There is general agreement that around two-thirds of the lambs lost die from two causes: first, abortion and stillbirths and, second, starvation and exposure (Figure 11.1). The moral here is that far more lambs die as a result of management failures than from infectious disease. High lamb losses are therefore not inevitable. It is estimated that at least half of the dead lambs could have been saved by taking the following measures.

- Reducing the number of weak lambs born by adequate feeding of the pregnant ewes.
- Early detection of starvation and hypothermia (chilling).
- Skilled lamb adoption techniques.

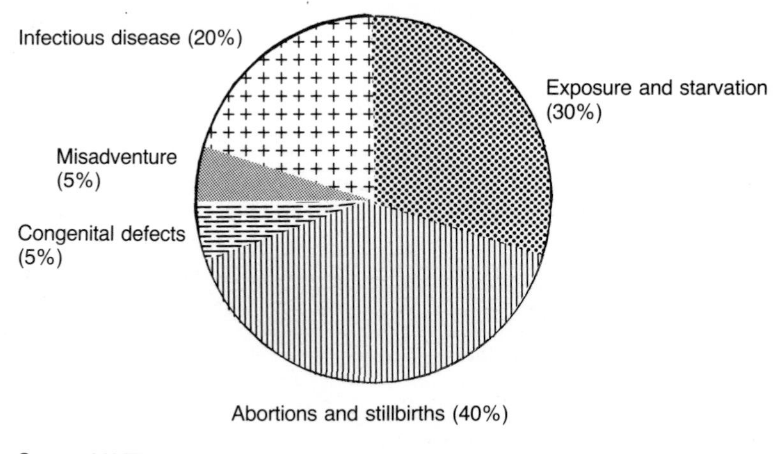

Figure 11.1 Causes of lamb deaths

These three measures are highlighted in the excellent MAFF Booklet 2525, *Lamb Survival*. They are dependent on improved stockmanship and involve little additional expenditure.

How can losses from the various causes be reduced?

Abortions and stillbirths

There are many causes of abortion. Enzootic abortion is one of the most important, and if this has been diagnosed all the ewes (including purchased replacements) should be vaccinated well before mating. Remember that pregnant women should not handle infected ewes because of the health risk involved. Toxoplasma may be more difficult to control because of the unavailability of a vaccine.

Bacterial causes such as *Campylobacter* or *Salmonella* are less frequent, but when they occur can cause severe outbreaks, and also infect humans – this emphasises the need for careful hygiene with any abortions.

When abortions occur, the aborted foetus and afterbirth should be retained for veterinary examination and diagnosis. Further aborted foetuses (following a further veterinary check) should be burnt, buried or put in a disposal pit, and the lambing pen should be disinfected and rested. The flock policy to prevent abortions should be discussed with your vet. Dystokia (difficult birth), mainly caused by over-large single lambs, is less common in prolific flocks. Its incidence is reduced by correct feeding of the ewe during pregnancy. This is aided by pregnancy scanning.

Starvation

Starvation is initially caused by a shortage of colostrum in the first few hours of life, and can only be avoided by close supervision at this time. Colostrum is vital to the lamb because it supplies energy and antibodies for protection against disease, and because it is laxative. Steps must be taken to ensure that all lambs take colostrum within 3 hours of birth, because soon after this the lamb's ability to absorb it is much reduced.

The ewe's udder should be checked as soon as she has lambed to ensure that her milk supply is adequate. If not, colostrum from another source must be administered to the lamb. The best choice is the surplus from ewes which have lost their lambs or from ewes with singles and plenty of milk. Otherwise, colostrum which has been kept in a deep freeze for up to 12 months may be used, the first choice being that from ewes, the second from goats, and the third from cows.

Colostrum should always be fed at body temperature. It may be fed from a bottle and teat. However, the use of a syringe and stomach tube has two advantages: firstly, it is quicker; secondly, following the use

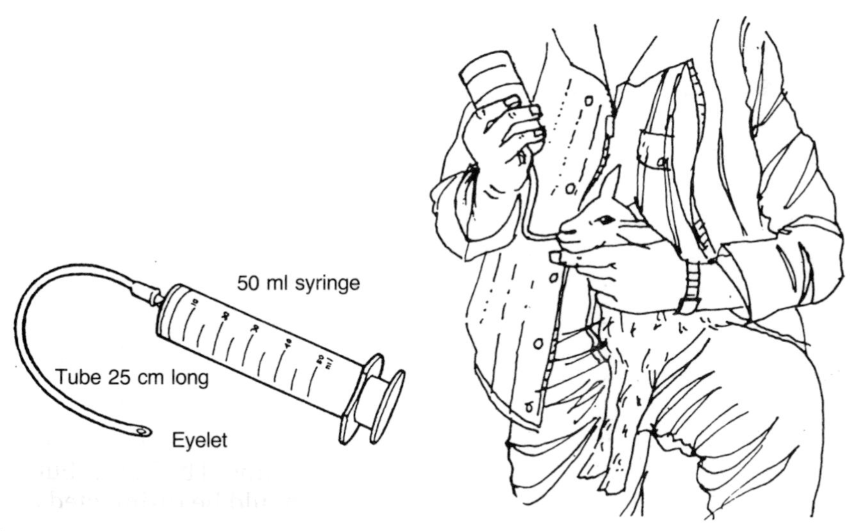

Source: MAFF Booklet 2525, *Lamb Survival*

Figure 11.2 Feeding a baby lamb by stomach tube

of this technique, lambs will suckle their mother or foster-mother more readily than following the use of a bottle and teat.

Feeding colostrum

- Thaw colostrum very gradually if frozen.
- Warm to blood heat. *Do not overheat.*
- Hold lamb as shown (Figure 11.2).
- Slide tube into side of mouth. *Do not force.*
- Slide down until 50–75 mm remain, or until resistance is felt.
- If lamb shows signs of stress remove tube and try again.
- When tube is in place, attach syringe and depress plunger slowly until empty (25 seconds).
- Leave tube in place, remove and refill syringe and repeat process until all feed has gone.
- Remove syringe and tube together.
- Wash and sterilise equipment. Keep in hypochlorite solution until required again.
- Do not use stomach tube in a weakly or moribund lamb as the tube may enter the windpipe. Colostrum in the lungs is fatal.

See also Table 11.1.

Table 11.1 Colostrum requirement to last lamb 4 hours

Weight of lamb (kg)	*Colostrum* (ml)
5.0	200
3.5	150
2.5	100

Source: Table and instructions above from MAFF Booklet 2525, *Lamb Survival*.

Exposure

Many lambs die as a result of hypothermia (chilling). The temperature of all weak lambs should be taken with a clinical thermometer in the rectum (Table 11.2). Where the temperature is sub-normal, the lamb should be dried, fed colostrum through a stomach tube and returned to the ewe. The temperature should be checked after half an hour.

Lambs less than 4 hours old and at severely sub-normal temperatures need warmth. They should be dried and then placed in a lamb warmer box kept at 40–45 °C until the lamb temperature reaches 38 °C, then

Table 11.2 Lamb temperature at birth

	(°C)
Normal	39–40
Sub-normal	37–39
Severely sub-normal	Below 37

Source: MAFF Booklet 2525, *Lamb Survival*.

fed colostrum through a tube and returned to the ewe. Again, the temperature should be checked after 30 minutes.

Severely hypothermic lambs over 5 hours old are a special case since they will have low blood glucose levels. They should be dried, injected with a 20 per cent glucose solution at 10 ml/kg weight and placed in the lamb warmer until revived. They can then be fed colostrum through the tube and returned to the mother. The glucose injection technique should be learnt under veterinary supervision.

Infectious diseases

The clostridial diseases include pulpy kidney, lamb dysentery, black disease, braxy and tetanus. We are very fortunate in having extremely effective clostridial vaccines available and some incorporate a *Pasteurella* vaccine. The necessary vaccinations of the ewes should be completed at least 2 weeks before lambing. These vaccines result in immunity to the clostridial diseases being passed on to the lambs in the colostrum. The immunity lasts for the first 10 weeks of the lamb's life.

Scour and watery mouth may also be troublesome. Their incidence may be reduced by good hygiene at lambing time, when both ewes and lambs are particularly vulnerable to infection. The veterinary surgeon can also help here.

Preparations for Lambing

Preparations for lambing fall under the headings of husbandry and equipment.

Husbandry

- Feed pregnant ewes correctly, particularly in the last 8 weeks before lambing.
- Check ewes' feet 8 weeks pre-lambing. Since all should have been

put in good order in the late autumn, or when housed, few should require treatment at this time.

- Complete vaccinations against clostridial diseases 2 weeks before lambing.
- Crutch the ewes 2 weeks before lambing.
- Pen early and late lambers separately. This is facilitated by the raddling of the tups at mating. The handling of ewes in late pregnancy should be minimised.

Equipment

Whether ewes are out- or in-wintered, lambing pens must be provided. They assist the mothering-up of ewes and lambs. They can also help to prevent the spread of disease, but only if they are carefully cleaned and disinfected between one ewe occupant and the next.

It is preferable to have the lambing pens in a building. If this is not possible, lambing pens erected outside should be in a sheltered position and covered. There should be one lambing pen per 8 ewes. The size should be 1.5 m^2 for average-sized ewes with twins. For large ewes or ewes with triplets the size should be 1.8 m^2.

At Rosemaund EHF, lambing pens constructed of metal hurdles have been completely satisfactory (see Plate 11.1). They are set up

Plate 11.1 Individual hurdle lambing pens

Table 11.3 Contents of the shepherd's cupboard

Lambing equipment	*Surgical equipment*
Soap Clean towels Buckets Disinfectant (non-irritant for handling sheep) Lubricating fluid Lambing rope Water heater or kettle	Prolapse equipment Rubber rings and applicator Foot-paring knife and secateurs
Survival kit	*Medicines*
Stomach tubes and 50 ml syringes Bottles and teats Colostrum in freezer (can be in handy-sized ice cubes) Tincture of iodine and plastic bottle Oral electrolyte solution Thermometer (and spare) Sterile syringes and needles	Worm drench Calcium borogluconate Magnesium sulphate solution (25 per cent) Glycerine Glucose (dextrose 40 per cent solution) Antiseptic pessaries Antibiotic (as prescribed by veterinary surgeon)
Miscellaneous	
Torch Record cards Weighing scales Ear tags and applicators Marking stick Aerosol colour sprays Drench bottle Disinfectant (general purpose)	

Source: Rosemaund EHF.

in an area of the sheep house which has a concrete floor, and this is recommended in order to facilitate disinfection.

The contents of a shepherd's cupboard are shown in Table 11.3.

A final point on preparations for lambing is that adequate trained labour must be available. It is a false economy to be short-staffed at this time. One experienced shepherd should be able to lamb approximately 250 ewes, with a part-time assistant.

Lambing the Ewe

Individual ewes may be moved from the group pen into a lambing pen just before or just after lambing. The lambing pens should be used in rotation, and should be cleaned, disinfected and deep-bedded after each occupant. It is advisable to have a few fresh pens in reserve for late lambers when disease risks are possibly greater. In the case of a severe disease outbreak, the pens should be dismantled, disinfected, and re-assembled on a clean site.

The supervision of lambing over the whole 24 hours is no problem in large flocks where the shepherds work a shift system. With small flocks, supervision may not be possible throughout the night, in which case the lights should be switched off for the period when the shepherd is not in attendance.

Most ewes lamb unattended and without difficulty. Difficult lambings are a small minority, mainly caused by large single lambs, and sometimes by over-fat ewes. An experienced shepherd will know when veterinary assistance is needed and will seek this at an early stage, realising that he cannot afford *not* to get professional help with difficult cases. Delay in taking the ewe to the veterinary surgeon may well lessen the chances of a successful lambing.

Lambing ewe lambs

Ewe lambs should be closely supervised at lambing. Parturition takes longer with ewe lambs than with mature ewes and the new-born lambs may take longer to suck (Table 11.4). It is a good idea to shear a little wool away from the ewe lamb's vulva and udder; this makes it easier for the shepherd to see if help is needed and for the lamb to find the teats.

It must be remembered that ewe lambs are still growing. To prevent a fall in performance in the following year, ewe lambs should receive

Table 11.4 Parturition time and time taken to suckle

	Parturition time (min)	*Delay from birth to suckling* (min)
Ewe lamb	87.5	70.5
2-year-old ewe	59.2	45.5
Adult ewe	50.6	33.5

Source: Rees, UCW, Aberystwyth.

supplementary feed after lambing until there is plenty of spring grass. Their lambs should be weaned as soon as possible, certainly by 12 weeks of age.

Mothering-up

The bond between the ewe and her lambs is formed in the first few hours after lambing. It depends mainly on the ewe's recognition of her own lamb(s) by her sense of smell. Therefore, if several ewes and lambs are penned together the ewes become confused because of the range of smells, and mis-mothering occurs, leading to lamb rejection and starvation, a main cause of lamb mortality.

The provision of lambing pens (particularly vital for ewe lambs), in which each ewe and her lambs are kept for 24–48 hours, is the best insurance against mis-mothering. This is a long enough period to allow the bond to be established. Some flockmasters use 'bulking up' pens for 20 or so ewes and their lambs after they leave the lambing pens and before turn-out to grass. This is not recommended (except where turn-out is delayed through bad weather) because it increases the danger of mis-mothering.

While in the lambing pen, the ewe should be offered 1 kg of good hay and 0.75 kg of high protein (18 per cent crude protein (CP)) compound/day in two feeds. Clean water must always be available.

Treatment of the New-Born Lamb

The availability of lambing pens facilitates the carrying out of several important tasks in the first 48 hours of the lamb's life. Castration and tailing using rubber rings should be carried out only after the lamb has had adequate colostrum.

- *Castration*. The earlier castration is carried out, the smaller is the check to the lamb's growth-rate. Rubber rings must be fitted *during the first week of life only* and this is a swift operation if the correct applicator is used. However, some discomfort is frequently caused for up to 15 minutes from the fitting of the ring, and during this period supervision is required to ensure that the lamb does not roll out of the pen under the hurdles. Alternative castration methods such as the use of the knife, searing iron or bloodless castrator can be carried out up to 3 months of age. Any castration at over 3 months must be carried out with the use of an anaesthetic.
- *Tailing*. Tailing is usually done by fitting a rubber ring *during the first*

week of life only. The short docking of tails at any age is forbidden by law. Sufficient tail must be left to cover the vulva of the female sheep and the anus of the male sheep.

- *Navel treatment*. Navel treatment is necessary to prevent infection via the navel cord, causing joint-ill and navel-ill, and should be carried out as soon after birth as possible. The authors have found nothing better than tincture of iodine sprayed on to the navel from a 'squeezy' plastic bottle. Iodine appears to speed the drying and shrivelling of the umbilical cord. It is also cheap.
- *Lamb identification*. Lamb identification is necessary for flock recording purposes and it helps to reduce mis-mothering. Poultry wing tags have been found to be perfectly satisfactory, with a loss rate of only 1 per cent at Rosemaund EHF up to the time of marketing the lambs. Spray-on aerosols have the advantage that they are clearly visible in the field for up to 2 weeks. If three colours are used, single lambs may be marked with a first colour, twin lambs with a second, and triplets with a third. If each ewe is sprayed with the same colour or number as her lamb(s), this is a great aid to the mothering-up of temporarily parted ewes and lambs in the field.

Treatment of the Newly Lambed Ewe

- *The udder*. The ewe must be checked for two healthy quarters giving a good secretion of milk.
- *The feet*. These should be checked and treated if necessary before turn-out.
- *Drenching against worms*. This is advisable unless the ewes have been previously drenched during pregnancy. The two alternative dates of dosing are compared in Chapter 10.

The triplet problem

Although the aim of the flockmaster is to breed two lambs from every ewe, it is inevitable in the prolific flock that some ewes will produce triplets or quadruplets. There are four alternative ways of dealing with triplet lambs: selling the third lamb, allowing the ewe to rear triplets, fostering the third lamb and artificially rearing the third lamb.

- *Selling the third lamb*. The sale of surplus lambs as soon as they have had their colostrum off-loads the task of their rearing on to someone else. There are always willing buyers of 'cade' lambs who require them to be fostered on to ewes which have lost their own lambs, or who wish to rear them artificially.

- *Allowing the ewe to rear triplets.* Until recently, the authors' experience has been that most ewes will rear only two good lambs. However, at Rosemaund EHF in early 1989, ten Mule or Dorset × Mule ewes rearing triplets were fed a ration which contained 10 per cent Megalac, a protected fat which is not broken down by the flora and fauna of the rumen. Consequently, the high energy ration is fully utilised by the ewe. The first year's results were very encouraging in that all 10 ewes reared strong lambs to weaning at 6 weeks. The ewe with triplets can be helped by giving her special treatment. Such ewes can be kept separate in a small parasite-clean paddock and fed extra concentrates, and a creep feed containing a coccidiostat should be made available to the lambs.
- *Fostering the third lamb.* Fostering is the most favoured solution of the triplet problem. One lamb (frequently the biggest) is fostered on to a ewe which has lost her lambs or on to a ewe which has a single and surplus milk. The first essential is to ensure that the fostered lamb gets its share of colostrum. There is a choice of fostering methods. Many flockmasters rub the lamb with afterbirth fluid from the adoptive ewe, or 'dress' the lamb in the skin of the adoptive ewe's dead lamb, provided this lamb did not die of disease. These methods ensure that the ewe gets the right smell. The alternative is to use a fostering box and a successful design is shown in Figure 11.3.

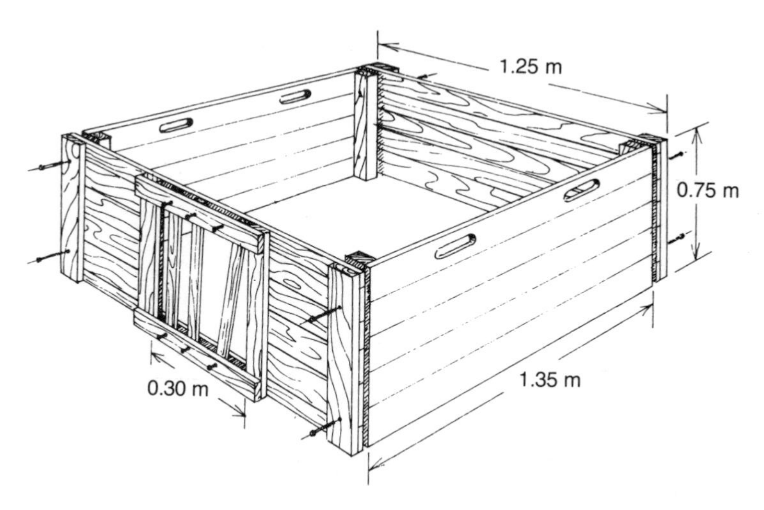

Figure 11.3 Fostering box

The ewe and lamb are placed in the box. The ewe is allowed food and water, but is fastened by the neck for 48 hours; this means that she is unable to turn and smell the lamb, which might lead to its immediate rejection. Meanwhile, the lamb has plenty of opportunity to suck. After 2 days the ewe is untied and left unrestrained in the box for a further 24 hours. The success rate with this fostering technique at Rosemaund EHF has been 80 per cent. Hygiene of the fostering box is important so that it is not a central focus for infection if disease occurs in the flock.

- *Artificially rearing the third lamb.* The bottle feeding of pet lambs is still a popular method of rearing triplet or orphan lambs, especially where the farmer's wife does the job and labour costs are not incurred. The alternative artificial rearing methods are now highly developed and very successful. However, they are costly.

  Following the feeding of colostrum, high-quality milk replacer is offered either rationed or ad lib. Compound feed (16–18 per cent CP) and water are on offer from 1 week old and weaning takes place at 28–30 days old. After weaning, lambs can be finished on a purchased complete diet concentrate or home mix compound. Alternatively, they can be finished at grass, but this *must* be free from parasites.

  This rearing system is technically sound but not always economically viable if full labour costs are included. It is fully described in ADAS Leaflet P3078, *Artificial Rearing of Lambs.*

Turn out to grass

The ewes and lambs should be turned out into small sheltered fields, with extra shelter provided in wild weather. Only strong lambs should be turned out and the best time of day for this is the morning, giving plenty of daylight hours for mothering-up. To aid the latter process it is also good practice to turn out no more than 10 ewes and their lambs to one field in a single day.

Where the sheep pastures are some way from the buildings, a livestock trailer can provide safe rapid transport. In a double-decker trailer, the ewes can be carried in the bottom and the lambs on the upper deck where they cannot be trampled under their mothers' feet.

Chapter 12

The Early Summer – Finishing and Marketing Lambs

Chapters 8–11 have covered the management of the ewe flock from the weaning of the lambs in the summer through to the spring of the following year. They have detailed preparations for mating, tupping management, pregnancy, lambing and the turn out of the flock to grass.

This chapter is concerned with the final period of the shepherd's year when all his efforts are brought to fruition. The requirement is for rapid lamb growth to bring as many lambs as possible to market before the summer fall in lamb values.

The severe fall in market price during the summer months has a profound effect on the returns from finished lambs.

Flock Management at Grass

Many details of the management of ewes and lambs at grass must be planned before turn-out. These have been covered in Chapter 5, but the authors believe that the following decisions are particularly crucial.

Fencing

Stock-proof fencing is essential to keep the flock at home. Inadequate fences result in frequent escapes and promote neither peace of mind nor good neighbourly relations. Sheep netting or electrified netting or three strands of wire provide suitable temporary fencing for the sub-division of a field into paddocks. However, there is a wide choice of secure boundary fences, and the specifications and costs of these can be obtained from ADAS.

Grassland

This is very much a case of horses for courses. On mixed farms (particularly those on light land) there is a good case for short-term leys.

Table 12.1 Weekly sheep guide prices in Great Britain in 1988

Week commencing	*p per kg dressed carcass weight (estimated or actual)*	*Week commencing*	*p per kg dressed carcass weight (estimated or actual)*
Jan 4	245.0	Jul 4	214.3
11	249.7	11	209.2
18	255.4	18	207.1
24	259.1	25	206.1
Feb 1	263.8	Aug 1	205.7
8	267.5	8	205.7
15	269.6	15	205.7
22	271.1	22	205.7
29	272.7	29	205.7
Mar 7	274.0	Sep 5	205.7
14	274.9	12	205.7
21	274.9	19	205.7
28	274.4	26	206.0
Apr 4	274.0	Oct 3	206.0
11	273.7	10	206.2
18	272.5	17	206.5
25	270.6	24	207.5
		31	209.1
May 2	268.4	Nov 7	210.7
9	266.1	14	213.8
16	262.5	21	216.8
23	251.2	28	219.9
30	244.4		
Jun 6	238.1	Dec 5	223.3
13	232.0	12	227.9
20	225.8	19	232.5
27	219.6	26	236.0

Source: MLC.

These may be ploughed out after 2 years and the accumulated fertility cashed in through the following cereal crops. Pure legume leys have been evaluated for this purpose and most have been found wanting. A case can be made for red clover as a 2-year ley and lucerne for 3–4 years, but there is a possibility that ewe fertility may be impaired if the sheep are flushed or tupped on these forages. Many years experience at Rosemaund EHF has shown them to be inferior to Italian ryegrass in yield and in reliability. The latter grass gives the flockmaster the following three advantages.

- Parasite-clean grass for one grazing year.
- The highest attainable forage yields, frequently 40 per cent higher than from long-term leys.
- Earlier spring growth than from longer leys.

The disadvantages are recurring reseeding costs, and additional fertiliser costs, especially for the nitrogenous fertiliser necessary to make up for the absence of clover in the sward.

Table 12.2 Price received for a lamb of 17 kg dressed carcass weight 1988

Month of sale	*Price* (£)
May	45.83
June	40.85
July	37.18
August	35.83

Source: Rosemaund EHF.

Certain tetraploid Italian ryegrasses such as Sabalan yield heavily in their first year and are very palatable. However, in their second year they may be outyielded by diploid varieties such as RVP. This variety, direct-sown at 35 kg/ha in August or September has given consistently high yields over 2 years at Rosemaund EHF.

On predominantly grassland farms, or where frequent reseeding is not regarded favourably, permanent grass or long-term leys will be the choice.

Unless there are special circumstances, e.g. on drought-prone land where cocksfoot comes into its own, perennial ryegrass/white clover leys which can last for 5–10 years have been found to be ideal for the ewe flock.

A mixture of late-flowering strains of perennial ryegrass and large-leaved white clovers has been very successful in providing leys giving high yields of nutritious forage with relatively low nitrogen inputs. For instance, at Rosemaund EHF results have been consistently good from the seed mixtures shown in Table 12.3 direct-sown in August or the first week of September.

Table 12.3 Long-term ley seeds mixture

	Seed (kg/ha)
Melle perennial ryegrass (diploid)	12
Meltra perennial ryegrass (tetraploid)	8
Milkanova white clover	4

Source: Rosemaund EHF.

Fertiliser application

On the basis of soil analyses, dressings of lime must be applied to ensure that soil pH remains at around 6.5 and certainly above 6.0. The phosphate and potash indices shown by the analyses should be 1 or 2, and to this end a maintenance dressing of 35 kg/ha of each has been applied annually at Rosemaund EHF. This application is carried out during the winter months when the ground is hard. Potash should never be applied during the spring because it enhances the danger of staggers.

Most swards grazed by lambing flocks should contain white clover, and the amount and pattern of nitrogen applications influence its persistence. Experience has shown that if the total annual dressing of nitrogen is limited to around 120 kg/ha the clover content can be retained.

As explained in Chapter 5, it is advisable to apply at least half the nitrogen in early and late season when the clover is not growing. The normal pattern followed at Rosemaund EHF has been to apply four dressings each of 30 kg/ha in early March, May, June and mid August. However, in recent years three dressings only, of 40 kg/ha each, have sufficed. These have been applied in early March, in June and in mid August. There is no evidence of a difference in response between three and four applications of nitrogen if the total dressing is the same. The spring application is in the form of ammonium nitrate, and the later dressings are of nitrogen–potash compounds in order to replenish the potash removed in conservation cuts.

Stocking rate

Successful flockmasters ensure high gross margin per hectare by maintaining high performance per ewe at optimum stocking rates. They do this by growing a lot of grass throughout the season, by stocking it to keep the grazing height correct (see Chapter 5) and by controlling parasitic worms.

In the grassier areas of the country overgrazing is quite rare. Even when it appears that there is little keep available, the sheep appear to be happy, leading flockmasters to appreciate the advantages of grazing comparatively short swards. The nutritional value is likely to be high and parasitic worm infestations may be lower than on longer grass because of the effects of sun and wind in desiccating the infective larvae. It has even been claimed that ewes are happy eating each day the grass which has grown during the previous night!

Undergrazing is a more frequent cause of poor performance. It is associated with the development of stemmy growth and seed heads, and this means a fall in 'D' value and in palatability. Where grazing numbers cannot be rapidly increased one solution is to increase the area of grass laid up for conservation. This is not easy where the flock is set stocked, but more readily achieved under paddock grazing or modified set-stocking regimes. Alternatively, the use of the topping machine is recommended. Some forage will be wasted, but the nutritive value of the sward is dramatically improved.

It is impossible to recommend stocking rates for all situations. Experience at Rosemaund EHF has taught us that on productive leys in the west of the UK a spring stocking rate of 22 ewes with their lambs/ha is possible. Where conservation cuts for hay or silage are taken, aftermath grazing becomes available during June and the stocking rate can then be reduced to 15/ha.

Grazing systems

The several alternatives are compared in Chapter 5.

Some flockmasters operate a paddock grazing system. They accept the additional costs involved in order to maximise the utilisation of their grassland, and in particular they are better able to conserve excess forage than those who opt for set stocking. A few go for the further sophistication of forward creep grazing because of the ability it gives the shepherd to finish and market lambs earlier. This is not in dispute, but the superior financial returns resulting from the better lamb sale pattern have not always justified this more complicated and expensive system of grazing management.

There is much to be said for opting for simplicity and just grazing three or four whole fields in rotation. This gives the opportunity to take occasional conservation cuts, and there are many examples of its successful operation.

The choice for the majority of flockmasters may lie between this simplified rotational grazing and the 'follow N' system.

Whatever the grazing system chosen, all leys must be closely defoliated in the autumn, otherwise winter kill can severely damage the sward, and this is particularly noticeable in the second winter of Italian ryegrass leys. Finally, the sheep must be removed to permanent pasture or to housing before they cause poaching damage in the winter. If poaching occurs, the sown species will be in part replaced by weed grasses, weeds and bare ground.

Feeding ewes at grass

There are two good reasons for feeding the ewe supplementary feed during the first few weeks at grass. The first is the necessity to stimulate milk yield upon which young lambs are so dependent in the first 6 weeks of life. The second is the need to prevent staggers (hypomagnesaemia) by feeding extra magnesium. The best way to ensure that each ewe receives her correct daily allowance is to incorporate it in a compound (see later in this chapter). Some flockmasters dislike feeding ewes at grass, considering that it increases the dangers of mis-mothering. However, the above two requirements appear to us to be overriding considerations.

The compound fed should have a crude protein (CP) level of at least 16 per cent, and at Rosemaund EHF the simple home mix shown in Table 12.4 has been satisfactory. To this, add sufficient calcined magnesite to ensure that each ewe receives 7 g/day. The even incorporation of the mineral/vitamin supplements is promoted by the rolling of the barley.

The feeding scheme shown in Table 12.5 is recommended. The ewes should receive compound (as a magnesium carrier) until the end of April, regardless of the amount of grass available. From May onwards, supplementary feeding of ewes is normally unnecessary since grass growth should be plentiful and, moreover, the ewe is past her lactation peak.

Creep feed for lambs

The creep feeding of lambs should be regarded as an aid to early marketing and therefore high sale value. It is also effective in reducing the grazing pressure which is particularly important in drier areas of the country where forage becomes short in mid season.

Creep feeder design is important. It is essential that where lambs enter a creep through an adjustable roller, the gap is regularly adjusted

Table 12.4 Compound mix for ewes at grass

	(kg/t)
Rolled barley	775
Soya bean meal	200
Sheep mineral/vitamin supplement	25

Source: Rosemaund EHF.

Table 12.5 Compound feed levels at grass

	(kg/ewe/day)
Weeks 1 and 2 at grass	0.75
Week 3 at grass	0.50
Weeks 4–6 at grass	0.25

Source: Rosemaund EHF.

Plate 12.1 Lamb creep feeder which excludes birds

to prevent bruising to the backs of the lambs in the later stages of finishing.

The feeder used at Rosemaund EHF (see Plate 12.1) is totally enclosed and successfully excludes birds. It is readily entered by lambs, which seem to regard it as shelter as well as a food store. The feeder is best sited close to the ewes' feed troughs so that when ewes come to feed the lambs follow and find their creep.

Creep feed can be home-mixed or purchased. The mix shown in Table 12.6 is used at Rosemaund EHF. For the first few days, a handful of flaked maize is added to aid palatability. On farms where coccidiosis is a recurring problem, a coccidiostat may be added to the creep feed.

It is worth providing creep feed from 10 days old, although no appreciable quantity is eaten until the lambs reach 3 weeks. At Rosemaund EHF the feed is initially offered in small amounts morning and night,

Table 12.6 Simple creep-feed mix for lambs

	(kg/t)
Rolled barley	775
Soya bean meal	200
Sheep mineral/vitamin supplement	25

Source: Rosemaund EHF.

with any left-overs being fed to the ewes. After 10 days, the troughs are replenished twice weekly so that feed is fresh and constantly on offer.

Where creep feed is provided it is worth continuing feeding it until the last lamb is finished off grass. Creep feeding may be discontinued when the decision is taken to store feed the remaining lambs.

Creep feeding is particularly advantageous where the stocking rate at grass is high and where aftermath grazing is unavailable because the ewes are wintered on straw and grass conservation cuts for hay or silage are not taken.

Dipping

It is at present a statutory regulation that all sheep must be dipped twice a year in order to control sheep scab. Although these dippings have no adverse effects on the growth of the lambs they can interrupt the weekly sale pattern. This is because lambs must not be sold within certain time limits after dipping, usually 10–14 days, depending on the dip used. At Rosemaund EHF, as many lambs as possible are sold prior to dipping. Even so, the first sale of lambs after dipping sometimes results in a few lambs being classified as overfat.

For detailed advice on the organisation and technique of dipping the reader is advised to read the ADAS Leaflet P2332, *Sheep: handling and dipping*.

Flock Health at Grass

Parasitic problems, metabolic disorders and trace-element deficiencies are all likely to be more prevalent under modern systems of intensive stocking. A year-round disease prevention programme is set out in Appendix B. The flockmaster must be on his guard against the following conditions.

Staggers (hypomagnesaemia)
The danger of staggers should be much reduced by the provision of additional magnesium in the ewes' diet for 6 weeks from the time of lambing. The mineral may be offered in molassed form to be licked from a ball in a trough or, alternatively, there are a number of palatable mineral supplements which may be left on free access. The problem here is that some ewes love them and some ignore them, and the latter are then at risk. The incorporation of calcined magnesite in the concentrate mix fed to ewes from lambing time is preferred because it ensures that each ewe gets her correct daily allowance. Ewes should receive 7 g calcined magnesite/head/day to provide about 4 g magnesium/head/day.

The other precaution is *never* to apply fertilisers containing potash to sheep pastures in the spring since this has the effect of lowering the magnesium content of the herbage.

Bloat
Bloat can occur when ewes graze all-clover swards. However, on mixed swards it is a rarity. At Rosemaund EHF, where the leys frequently contain 10–20 per cent white clover, bloat has not been seen in sheep over the last 25 years.

There are two relevant factors here. Firstly, the flock is turned out after lambing in late March, whereas clover growth is negligible before the month of May; this means that clover is introduced gradually into the ewes' diet. Secondly, the sudden introduction of ewes to a clovery sward which can occur with paddock grazing is less likely under the set stocking or modified paddock-grazing systems now prevailing.

Parasitic worms
The combination of clean grazing and drenching is a proven method of worm control fully described in Chapter 5. It should be adopted wherever possible. However, in many farm situations (see Chapter 5) it is a non-starter. For this reason, alternative control measures have

been developed at Rosemaund EHF. These have resulted in long-term leys being grazed for up to seven consecutive seasons with no decline in flock performance caused by parasitic worms. The routine is as follows.

- Dose all ewes when housed at the turn of the year.
- Dose all lambs when 4 weeks old and then at 3–4 weekly intervals until sale. Always dose lambs when they are moved on to aftermath grazing or other relatively parasite-clean grazing. The other very important time to drench is at weaning, as this stress can allow worms to get the upper hand.

There are several relevant points concerning the choice and use of anthelmintic drenches.

- When selecting drenches during the lamb-selling period, choose those with short withdrawal periods. This imposes less restriction during the lamb-selling period.
- There are reports of the development of worm resistance to some anthelmintics and this should be borne in mind and discussed with your veterinary surgeon.
- Worm lambs every 3–4 weeks as a routine. This can be brought forward or back a few days to coincide with changing pasture. In the drought year of 1976, lambs at Rosemaund EHF grew well and did not scour. The worming routine was therefore abandoned. In the following year, lambs were lost from *Nematodirus* infection. The moral is that regular dosing keeps down the level of worm infestation and so protects the following year's lamb crop.

It is also worth remembering that fields not grazed by lambs during the April and May of the previous year will pose less risk from *Nematodirus* worms. It is useful to bear this in mind when planning silage and hay cuts.

The worm control measures described were developed with veterinary advice. Such advice to suit his own circumstances should be sought by every flockmaster.

Coccidiosis

As with all diseases, prevention must be better than cure, and preventative measures include the following.

- Feeding a coccidiostat in the lamb creep feed. The effectiveness of this depends on the level of creep feed intake. In a 2-year experiment at Rosemaund EHF, lambs fed creep feed plus coccidiostat had better live-weight gains and an earlier average sale date than lambs fed creep feed with no coccidiostat. The 2 years of the trial were 1987 and 1988 when there were no clinical signs of coccidiosis in either group, so it was concluded that the coccidiostat had controlled sub-clinical coccidiosis.

- The inclusion of a coccidiostat in the concentrate fed to ewes in flocks where lambs are not creep fed. Experience shows that this does not eliminate the disease but it does reduce the infestation of pastures with oocysts. This in turn means that grazing lambs are infected at a lower level.

The symptoms of coccidiosis usually occur at 4–8 weeks old in late April/early May. It is most important to watch carefully from this time of year for the characteristic chocolate-coloured bloodstained faeces. At the first sign of these *the whole flock* must be treated, since coccidiosis must be regarded as a flock problem rather than an individual disease.

A number of injections and drenches are available for treatment. At Rosemaund EHF, drenching with sulphonamide on 3 successive days has always been effective in clearing up the problem.

Nematodirus is frequently also a danger at this time and because of the importance of differential diagnosis it is strongly recommended that preventative and curative measures should be discussed with the farm's veterinary surgeon.

Drinking water must be available to lambs as well as to ewes. Lambs eating creep feed are likely to require water especially during dry periods, and any diarrhoea caused by *Nematodirus* infection or coccidiosis will cause dehydration which may result in kidney damage. At Rosemaund EHF, low drinking troughs for lambs were installed some 15 years ago following cases of nephrosis after an outbreak of coccidiosis.

Cobalt deficiency ('pine')

Cobalt is lacking in some soils, including the old red sandstones. Cobalt is essential for the synthesis of vitamin B_{12} within the rumen of the sheep. The deficiency is characterised by the lambs losing weight and condition after weaning even when grazing an abundance of grass.

It has been proved that 'pine' can be prevented by spraying 750 g cobalt sulphate/ha in 180 litres of water on to the sheep pasture each year in February or on to the aftermaths immediately after taking cuts for silage. Alternatively, cobalt deficiency can be rectified by administering a cobalt bullet orally or by the injection of vitamin B_{12}. These three treatments are all effective and their current costs should be compared when choosing between them.

The importance of cobalt was shown in an experiment conducted at Rosemaund EHF in 1977. Three groups of 20 lambs were grazed on pasture which had not been sprayed with cobalt sulphate. The lambs in one group were treated with cobalt bullets, those in the second group had monthly injections of long-acting vitamin B_{12} and the lambs in the third group were left as a control with no treatment. At the start of the experiment on 29 July 1977 the mean weight of each group was 29.4 kg. The experiment was concluded on 3 October 1977 when lambs which had received either a cobalt bullet or vitamin B_{12} injections were marketed at a mean weight of 37 and 38 kg, respectively. The control lambs had lost weight and weighed only 27 kg at the final date.

The control lambs showed symptoms of cobalt deficiency, exhibiting poor general condition and coats, and matted facial wool caused by running eyes. To rectify the condition these lambs were injected with vitamin B_{12} and they immediately started to improve in condition and live weight. The severity of cobalt deficiency varies from year to year and responses of this magnitude will not always be obtained. In areas with cobalt-deficient soils, routine spraying can be recommended as an effective preventive measure, but it must be remembered that all cases of unthrifty lambs are not the result of cobalt deficiency.

Selenium deficiency (white muscle disease)

White muscle disease is caused by a deficiency of selenium in combination with low levels of vitamin E. Selenium deficiency in sheep is caused by an inadequate intake of this trace element. If the animal's foodstuff is grown on soils deficient in selenium, the dietary intake will be low unless supplemented by some means.

The symptoms of the disease vary, depending on which muscles are affected, but lambs with white muscle disease usually have difficulty in walking, their hind legs are often held in a characteristically stiff manner and they lie down to rest at frequent intervals. If the heart muscles are affected, pneumonia or sudden death may occur. Most cases are seen in lambs between 4–8 weeks old.

White muscle disease is always a flock problem and the vet should be consulted immediately symptoms are observed. Affected animals can be treated with a combined selenium and vitamin E injection, but experience has shown that some lambs respond to this treatment and others do not.

To prevent other lambs going down with the disease they should be given selenium supplementation immediately. This may be in the form of a proprietary drench or an injection. Care must be taken not to over-dose the animals because selenium poisoning can occur, and veterinary instructions must be followed carefully.

Long-term prevention depends upon formulations of the ewe's ration to ensure that it contains adequate amounts of selenium.

White muscle disease was first diagnosed at Rosemaund EHF in 1966 and it became a real problem in 1975 when high-quality silage was fed to the pregnant ewes. This required little compound supplementation and, hence, the ewes received minimal amounts of minerals.

The disease was studied in detail in close collaboration with Dr Peter Anderson of the MAFF Central Veterinary Laboratory, and it was discovered that the soils at Rosemaund EHF were low in selenium and, in consequence, so was the herbage. Because of the results obtained from these experiments, many feed compounders and manufacturers of mineral supplements adjusted the formulation of their products to include the correct amount of selenium, and this has largely overcome the problem.

There are, however, still special cases of farms where white muscle disease occurs despite the feeding of appropriate supplements, and here the ewes can be injected with a long-acting preparation of barium selenate. This treatment may keep selenium levels satisfactorily high for up to 2 years, but it should be used only on veterinary advice.

Copper deficiency (swayback)

Copper deficiency can be important in areas where pastures are high in molybdenum, which reduces copper absorption. Pregnant ewes should have an adequate copper status to ensure the nervous system of foetal lambs is properly developed. Copper-deficient ewes can give birth to 'swayback' lambs. These are either born unable to stand or show hind-leg weakness from 3 weeks old.

Ewes out-wintered may eat appreciable amounts of soil with the grass during February and March and this soil will also lessen copper absorption. During February each year the Ministry of Agriculture forecasts the likely incidence of swayback (and twin lamb disease) based on the amount of snow. Farmers feed more concentrates during frosty weather and snow cover – less soil is then eaten and the absorption of copper from concentrates is good.

Ewes can be blood-tested in early or mid pregnancy to check their copper status and, if low, this can be remedied by copper injections or copper capsules given orally on veterinary advice.

Ewes housed for more than 4 weeks before lambing are unlikely to have any copper deficiency problem. A danger exists of copper poisoning in housed ewes and growing lambs. The absorption of copper from concentrates is very good even though there is no added copper in sheep concentrates or sheep minerals. Copper can accumulate in the liver until a sudden crisis occurs. The animal becomes jaundiced and dies.

Copper deficiency can also result in lack of growth in older lambs on some farms. The wool is steely in appearance and the lambs may succumb more readily to other diseases. Blood samples can be taken to test the lambs' copper status.

Lamb Marketing

It is impossible to over-emphasise the importance of this subject. Lamb competes for the housewife's favour with beef, veal, pork, chicken, turkey and other meats. It will continue to decline in popularity unless much more effort is put into every aspect of marketing. Particular improvements required are in its leanness, uniformity, presentation and continuity of supply.

For much of the information on marketing in this chapter we are indebted to the MLC and to Alan Barrett, of Barrett and Baird Ltd, who is a leading authority on marketing lamb.

The EC sheep regime

The EC sheep regime has had a favourable influence on marketing in three ways. Firstly, the variable premium on lambs for slaughter and the annual payments made on breeding ewes have brought stability into sheep production. Secondly, the exclusion of fatter carcasses from the scheme has encouraged the marketing of leaner lambs. Thirdly, the seasonal guide-price pattern has encouraged a more even spread of marketing through the year, and therefore better continuity of supply.

In recent years, buyers of lamb have defined more precisely their requirements in terms of carcass weight, fatness and conformation.

Carcass fatness

Regardless of carcass weight, the market preference is for lean lambs classified 2 or 3L (Table 12.7). Lambs graded 3H are acceptable in a few markets, but 4L carcasses are normally penalised on price, and the fattest carcasses (4H and 5) do not attract the variable premium. In spite of the often expressed market abhorrence of fat, abattoirs still report that between one-quarter and one-third of lamb carcasses are classified 3H or fatter, with the national average for these over-fat lambs in 1986 being 27.6 per cent.

There is no doubt that fatness is directly related to yield of saleable meat (the fatter the carcass the lower the yield), and in view of the low value of fat, the figures in Table 12.7 explain the buyer's reluctance to purchase lambs in fat classes 4 and 5.

In recent years, both the Committee on Medical Aspects of Food Policy (COMA) and the National Advisory Committee on Food Education (NACNE) have recommended a reduction in the total fat intake in our diet, particularly if this fat is in the saturated form. Since many lamb joints may contain a higher proportion of fat than beef, poultry or pork, these recommendations have not helped the sale of lamb.

Table 12.7 Composition of carcasses in MLC fat classes

	Fat class						
	1 (%)	2 (%)	*3L* (%)	*3H* (%)	*4L* (%)	*4H* (%)	5 (%)
Saleable meat	95.0	93.0	92.0	91.0	90.0	89.0	85.5
Fat trim	4.0	6.0	7.0	8.0	9.0	10.0	13.5
Other trim	1.0	1.0	1.0	1.0	1.0	1.0	1.0
Lean	67.5	64.0	62.0	60.0	59.0	57.0	56.0
Fat	10.5	16.0	18.5	21.0	24.0	26.0	29.0
Bone	22.0	20.0	19.5	19.0	18.0	16.0	15.0

Source: MLC.

It is remarkable (and unfortunate) that such an important carcass assessment as fat content is made by subjective means. However, a number of objective methods which rely less on human judgement are now under development. These include video image analysis using a video camera to assess visually the proportions of lean and fat, and automatic probes as used to assess fat thickness in pig carcasses.

A third system is ultrasound, the early results from which were described to the British Society of Animal Production in 1986 by A. V. Fisher and S. J. Page of the Food Research Institute, Bristol Laboratory. At 37 °C, the speed of ultrasound through muscle is about 10 per cent greater than through fat tissue, and this difference can be used to measure their proportions in a carcass. The researchers identified two suitable sites in sheep carcasses, and concluded that the system gives measurements potentially useful in objective carcass classification. With presently available apparatus, the accuracy of prediction is greater in beef than in sheep carcasses. However, the importance of accurate objective lean and fat assessments in both live animals and carcasses is such that these developments are being pursued at research centres, on experimental husbandry farms and in abattoirs. In the latter, modern slaughter line systems operating at 450 lambs/h pose a problem in the speed of collection of this information.

There is also an associated need for electronic data capture systems to be used in abattoirs to bring together objective information on individual animals in terms of weight and carcass quality; this would be of great value both to the flockmaster and to the meat buyer. Successful trials in this area have been conducted at the Barrett and Baird abattoir and elsewhere. However, a thorough comparison of the available schemes and a decision as to the best system is needed before such is sanctioned for use nationally.

Carcass conformation

Conformation is the overall shape of the carcass based on a visual judgement. There is little evidence that conformation is a good guide to meat yield or to the proportion of the higher-priced joints. Therefore, it has not had a high priority in the efforts of sheep breeders. They have preferred to concentrate on improving the prolificacy of their flocks, and to a lesser extent the milking ability. In addition, the earlier marketing of light-weight lambs brought about by the intense dislike of fat has meant an apparent deterioration in conformation; this is because increased fat cover can improve the shape of the carcass and give the impression of good conformation!

Conformation is now of increasing importance in the marketing of lambs. Carcasses are classified on the scale EUROP with E indicating the best conformation and P the poorest. Some important supermarket chains pay a premium for carcasses classified E or U and those classified O are less easy to sell. In addition, Alan Barrett reports that overseas buyers are increasingly critical of lambs with poor leg shape and inadequate eye muscle in the loin. All meat is selected by the consumer on its appearance and, if it looks right, the public will buy it.

Some hill flocks selling well under 50 per cent of their lambs classified R or better have a big problem, and unfortunately there is little which can be done to improve conformation by improved management. Only better breeding stock will give better-muscled lambs, and we must look to meticulous selection of the most blocky shaped rams to be used as terminal sires, whether from the best British or continental breeds.

Carcass weight

The main demand in the UK and elsewhere in northwest Europe is for carcasses in the range 15–20 kg. There is a trend to lighter carcasses in the summer, and to heavier ones in the winter. There are also regional differences of preference within the UK. In 1986, the average lamb carcass weight in the UK was 17.3 kg, with higher average weights of just over 18 kg in Scotland and in the southeast and southwest of England, and the lightest average weight of 15.6 kg in Wales. Carcasses for export average 16–19 kg to France, 12–15 kg to the Middle East and 11–12 kg to Mediterranean countries.

There is a small but developing market for heavy carcasses in the range 22–30 kg. The disadvantages of taking lambs to these high weights are, to the flockmaster, a lowered stocking rate and to the butcher, perhaps too much fat to trim. However, enthusiasts point out that with the right breeding and feeding, large does not necessarily mean fat, and a few supermarket chains are regular buyers of large lean lambs.

Several new methods of butchering were first demonstrated by the MLC in 1982 and, although developed for a wide weight range of lean lambs, these would seem particularly appropriate for large carcasses destined for the catering trade. The new cutting techniques include the production of lamb slices and steaks, and a range of boneless joints. Because they are boneless, such joints are more expensive per kilogram, but the MLC points out that boneless cuts are attractive, convenient in size, easy to cook and easy to carve, leaving no wastage on the plate.

Sheep carcass classification

Classification provides a simple description of a carcass in terms of category (lamb or hogget), weight, fatness and conformation. It is carried out by MLC staff in the abattoir. At present it is subject to human error but, as mentioned previously in this chapter, there are high hopes that equipment may become available to allow objective assessments of fatness.

Both producers and the meat trade gain much from carcass classification. The flockmaster receives information on the types of carcasses he is producing, and can compare these with the trade's stated preferences in order to adjust his breeding and/or husbandry and marketing practices. The meat buyer can use classification to define his purchasing specification. A further important use of the scheme is as a description of carcasses to be used in the definition of premium and penalty-pricing schemes.

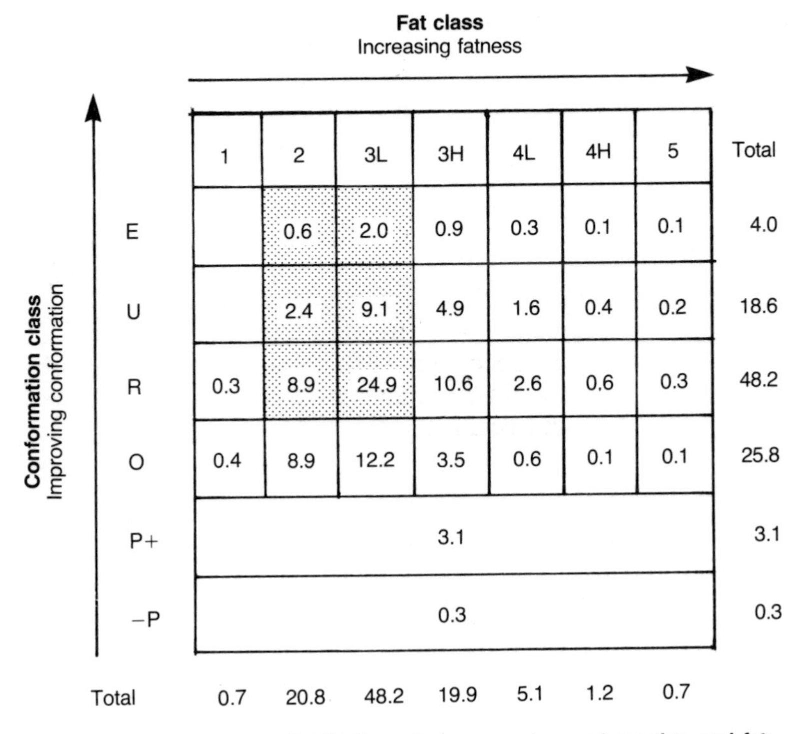

	1	2	3L	3H	4L	4H	5	Total
E		0.6	2.0	0.9	0.3	0.1	0.1	4.0
U		2.4	9.1	4.9	1.6	0.4	0.2	18.6
R	0.3	8.9	24.9	10.6	2.6	0.6	0.3	48.2
O	0.4	8.9	12.2	3.5	0.6	0.1	0.1	25.8
P+				3.1				3.1
−P				0.3				0.3
Total	0.7	20.8	48.2	19.9	5.1	1.2	0.7	

Figure 12.1 Percentage distribution of carcasses by conformation and fat class

The target area for producers in order to satisfy most buyers (shaded in Figure 12.1) is within fat classes 2 and 3L and conformation classes E, U and R. In Figure 12.1, 47.9 per cent of carcasses were within this preferred specification.

Carcass specification

With the gradual replacement of small butchers by the supermarkets and large multiple chains, the requirements of meat buyers are becoming ever more precise. As Alan Barrett has pointed out, 'It is no longer a case of "I have produced it so you have got to eat it".' The big buyers are interested only in carcasses within their tight specifications, and will go elsewhere (including abroad in some cases) if the right lambs are unavailable. Future developments in packaging could mean that, even in the chill condition (not frozen), meat will be transported successfully over long distances due to greatly enhanced shelf life. The moral is obvious. Home-produced lamb must be made available in quantity and to a tight specification through as much of the year as possible.

Table 12.8 details the 1988 specifications for home-produced lamb of five of the biggest supermarkets.

All the supermarkets specified short, thick, well-fleshed legs and well-fleshed loins. Fat should be white or creamy white, and the fat cover over the loin eye muscle not over 6 mm in thickness (not over 3 mm in one specification).

The buyers would not accept carcasses showing:

- Bruising.
- Abscesses (often caused by faulty injection technique).
- Visual contamination, including blood or excreta stains.
- Evidence of chemical residue.
- Taints or odours.

The ram lamb controversy

This subject has been included in the marketing section because ram lambs stand or fall depending on their market acceptability.

Whether or not to castrate male lambs has been an issue for many years. There is an underlying feeling in the meat trade that the meat from entire male lambs has a taint which the public can detect and will avoid. It is also claimed that the carcasses of entire lambs are more difficult to skin.

At Redesdale EHF during the period 1983–5, an experiment was conducted to investigate the advantages and disadvantages of keeping the males entire. Sets of twins (150 in all) either Suffolk × Mules or Dorset Down × Mules were selected at birth for the trial. One lamb from each set of twins was castrated and the other left entire. The lambs grazed on the hill with their dams until they were weaned. They were selected for slaughter by weight and condition. The results are given in Table 12.9. Table 12.9 shows that the entire lambs had a 7 per cent faster growth rate than their castrated twin brothers.

Despite a lower killing out percentage (Table 12.10) the entire lambs yielded an extra 0.5 kg of carcass weight and took 2 days less to finish. In each of the 3 years, the entire lambs had slightly leaner carcasses.

In 1983, the first year of the experiment, 62 carcasses were examined by the Food Research Institute (FRI) laboratory at Bristol. The laboratory dissection of the carcasses showed that the entire lambs had comparatively large shoulders and were generally leaner

Table 12.8 Home-produced lamb specifications of five buyers

	Buyer				
	1	*2*	*3*	*4*	*5*
Breed			Unspecified by all		
Age/Category	Wether or gimmer lambs	Under 12 months	—	Lambs or hoggets	Ram lambs under 5 months Wether lambs under 8 months
Carcass weight (kg) lambs	15.0–20.5	15.4–17.7	15.0–19.0	14.5–17.2 (England & Wales) 15.5–20.8 (Scotland)	16.0–20.0
Carcass weight (kg) hoggets	15.0–20.5	16.3–19.1	—	—	—
Carcass classifications	2 3L 3H	—	2E 2U 3LE 3LU	2E 2U 2R 3LE 3LU 3LR	1E 2E 3LE

Sources: Various supermarket buyers.

Table 12.9 Lamb performance

	Birth weight (kg)	*Weaning weight* (kg)	*Sale weight* (kg)	*Daily live-weight gain birth to slaughter* (g)
Entire lambs	5.2	37.7	43.7	306
Castrate lambs	5.1	36.5	41.4	286

Source: Redesdale EHF.

Table 12.10 Carcass details

	Days to sale	*Carcass weight* (kg)	*Killing out* (%)	*Carcass classification* 2 (%)	3L (%)	3H (%)	4 (%)
Entire lambs	129	18.8	43.0	27	63	7	2
Castrate lambs	131	18.3	44.2	22	64	9	4

Source: Redesdale EHF.

and contained less kidney knob and channel fat than the carcasses of the castrated lambs. The FRI taste panel could not detect any of the so called 'male taint' and found the carcasses of the entire lambs similar in flavour, tenderness and juiciness to those of the castrated lambs.

To summarise, most experimental evidence shows that ram lambs grow faster than wether lambs and produce slightly leaner carcasses, similar in other respects to those from castrated males. Nevertheless, it is advisable to check their acceptability in the chosen markets beforehand. Some buyers specifically state that they purchase only wethers and gimmers. If the lambs are left entire they should be finished by 5 months old because of their impending sexual maturity.

Selection for slaughter

Skilled selection will optimise returns by providing carcasses which attract premium payments. Remember that earlier maturing breeds and crosses lay down fat at lighter weights than later maturing sheep, and likewise that ewe lambs mature earlier than wether lambs which themselves mature before ram lambs. It follows that ewe lambs of an early maturing type can give an 'early warning' of the sale period, and from this time onwards it is most important to select weekly for sale. If this is neglected late sales will result, with the following disadvantages.

- Possible additional lamb losses through disease or accident.
- Shortage of feed through over-stocking of grassland.
- Additional costs of supplementary feed and worm drenches.
- Lower lamb sale value because of over-fatness or declining market price.

How, then, should lambs be selected for sale? It is clear that this should be on the basis of degree of finish judged by handling the live lambs. Selection by weight is a very poor indication of marketability, and selection by age is even worse.

How to assess fatness

This section is reproduced from the excellent MLC Booklet *Lamb Carcass Production: planning to meet your market.*

Practice in handling live lambs to assess individual fatness levels is essential, and, until experience is acquired, carcasses from each batch sent for slaughter should be examined at the abattoir. If a visit to the abattoir is not possible then carcass classification results can be used to show how accurate the selection for fatness has been.

There are four points on a lamb where handling will provide a reliable guide to the fatness of its carcass. The two most important points are A and B, but if in doubt C and D can be used as an additional guide (Figure 12.2).

The level of fatness at A is assessed by handling the tail root to see how much fat is covering the individual bones of the tail. As lambs become fatter, it is more difficult to detect individual bones.

Fatness at B is assessed by placing the hand over the spinous and transverse processes of the loin to assess their prominence. The less prominent the processes the fatter the lamb.

Based on the handling at A and B, an overall score for fatness from 1 (leanest) to 5 (fattest) can be made – similar to the fat scale in the MLC Sheep Carcass Classification Scheme (Table 12.11).

The handling technique can be used to assess the fatness of all breeds, but it is important to allow for differences in thickness of wool. It is also important to apply minimum pressure with the fingers to avoid bruising which can seriously lower the financial value of the carcass.

A Around the dock

B Along the spinous processes of the back bone and over the eye muscle and the tips of the transverse processes in the lumbar region

C Over the ribs

D Along the spinous processes of the back bone over the shoulder

Figure 12.2 Handling points

Uniformity and continuity in marketing

With fewer buyers each purchasing large numbers of lambs for supermarkets and for export, today's emphasis is more and more on uniformity. These buyers have little interest in mixed groups and require batches of lambs even in weight and finish. Although the flockmaster may be handicapped by the possession of several breeds and crosses, attention to the presentation in the market of clean, even groups of lambs will always pay.

The trade also requires continuity of supply on a year-round basis rather than the seasonal marketing of lambs. We can reorganise to satisfy this demand by breeding some of our ewes out of season as described in Chapter 7. There is already a well-established market for April-finished Easter lamb from flocks lambing in December

Table 12.11 The assessment of fatness by handling

Fat class	*Dock*	*Loin*
1	Fat cover very thin. Individual bones very easy to detect	Spinous processes – very prominent. Individual processes felt very easily. Transverse processes – prominent. Very easy to feel between each process
2	Fat cover thin. Individual bones detected easily with light pressure	Spinous processes – prominent. Each process is felt easily. Transverse processes – each process felt easily
3	Individual bones detected with light pressure	Spinous and transverse processes – tips rounded. With light pressure individual bones felt as corrugations
4	Fat cover quite thick. Individual bones detected only with firm pressure	Spinous processes – tips of individual bones felt as corrugations with moderate pressure. Transverse processes – tips detected only with firm pressure
5	Fat cover thick. Individual bones cannot be detected even with firm pressure	Spinous and transverse processes – individual bones cannot be detected even with firm pressure

Source: MLC.

and January, and these lambs are now attracting interest for export. Seasonal price differentials and husbandry developments (see Chapter 6) have also encouraged the finishing of more hoggets from the New Year through to March. The development of sheep systems which will allow UK lambs to be marketed throughout the year is being actively pursued on the Ministry's experimental husbandry farms.

Carcass damage

Carcasses may be devalued by the presence of abscesses, by bruising and by dirty fleeces.

It has been shown by MLC surveys that almost 1 per cent of lamb carcasses have abscesses, and it is calculated that this represents a loss of approximately £5 million/year in value of carcasses and hides; it also harms export prospects. The problem is caused in part by injection in the wrong site such as the shoulder or leg whereas the correct place is the upper neck. Another cause is inadequate hygiene when the needle is not disinfected between injections, or when lambs are injected through wet or dirty fleeces.

Bruising can be avoided by making sure there are no sharp projections in pens and handling units, by banning the use of sticks and of dogs which bite, and by treating sheep in a quiet manner. They should never be lifted by the fleece (which causes wool pulling and bruising) but rather by the neck and flank, and particular care should be taken when sheep are being loaded on to vehicles. Hygiene in the abattoir is assisted if the animals are regularly dagged to reduce the number of dirty fleeces. It is advantageous if hoggets to be folded on roots have the under belly shorn first. This is now becoming a more common practice.

Many of the above points play a part in improving animal welfare. Likewise, the abandonment of ear punching of lambs accepted for variable premium payments and its replacement by the use of suitable eartags would be another improvement in the humane treatment of sheep.

Methods of marketing

The choice is between live-weight and dead-weight marketing. Both have advantages from the producer's point of view and both will continue to operate as alternatives between which he must decide. At present, about 70 per cent of lambs are marketed live-weight. The principal advantage here (apart from the social attractions of the live market!) is that dissatisfied producers can refuse to accept the highest bid and take their lambs home. They can then be re-presented at the market in more suitable condition and hopefully on a day when prices are firmer.

Dead-weight marketing has two main advantages. The first is that it undoubtedly allows more accurate classification since the grader views the carcass. The second is the availability of feedback on carcass quality. To obtain the latter, it is important to set up a good relationship with the abattoir, and to follow through occasional groups of lambs and view their carcasses. On these occasions it can be rewarding to ask the grader to demonstrate the carcass points for which he is looking.

The information on lamb quality obtainable from dead-weight

centres can be the basis for future decisions on the breeding and feeding of the flock, and gives guidance valuable in the selection of lambs for sale. Lambs which are eartagged can be traced back to individual rams in order to assess the influence of particular sires on carcass quality.

Several surveys have compared returns from live-weight and dead-weight marketing. Firm conclusions cannot be drawn, but certainly dead-weight sale is likely to be advantageous compared to sale in a live auction where large deductions are made from the half weight of the lamb. This occurred when Dr Mike Tempest compared three groups of lambs marketed by the Harper Adams College of Agriculture and found a £4–5/lamb advantage to the dead-weight system.

Regardless of the method of sale there is evidence that long or indirect journeys from the farm to the market or abattoir can reduce not only the live weight, but the carcass weight also. Some studies have found carcass losses of 0.5–1.0 kg/lamb where fasting exceeded 24 hours. In MLC trials carried out in 1984/5, D. G. Evans *et al.* used 9 batches of lambs to compare the effects of a long indirect route, taking 26 hours from farm to abattoir, with those of a short direct route, taking only 5 hours. The long route lambs lost an extra 2.54 kg of live weight and an extra 0.47 kg of carcass weight, and their killing-out percentage was the poorer by 1.1 per cent.

Quality lamb groups

Quality lamb groups are marketing co-operatives whose producer members agree to contract all or some of their lambs to the group. The groups agree to supply abattoirs with numbers of lambs of agreed quality over defined periods of time. In return for a commitment to market quality lambs via a group, a producer is paid a premium. The concept of these groups is a good one but not all have proved successful. In some cases the overheads of the groups have been high and in consequence the premiums paid to the members have proved insufficient to ensure their continued loyalty.

Quality lamb groups may gain in popularity if the variable premium scheme disappears. In this event, producers might wish to use the services of proven fieldsmen in order to improve the efficiency of selection of lambs for sale.

Summary

Lamb carcass quality has an increasing influence on the profitability of the ewe flock. The image of lamb in the eyes of the housewife is of a tender tasty meat which is, however, too fatty and wasteful, difficult to carve and lacking in versatility. The last two criticisms may be met by the new cutting techniques, which could satisfy the demand for boneless joints and increase the attractions of lamb for the catering trade. The flockmaster has his part to play in marketing uniform undamaged carcasses which meet modern specifications, and in improving continuity of supply.

The present target for most home and export markets is the 15–20 kg carcass in fat classes 2 and 3L and conformation classes E, U and R.

PART IV

The Profitable Ewe Flock

Chapter 13

How to Improve Flock Performance

The purpose of this final chapter is to pinpoint the key husbandry factors which make for the physical and financial success of the sheep enterprise. These are then combined to form a management blueprint for a lowland ewe flock to achieve gross margins which at present prices can exceed £800/ha.

The Need for Accurate Flock Recording

Physical and financial targets should be set for every flock each year. Both ADAS and the MLC publish suitable targets annually or, alternatively, the flockmaster may wish to set his own, perhaps after consultation with his local adviser. At the end of the sheep year when the lamb crop has been sold or valued-out, the results achieved should be compared with the targets set. This will reveal any deficiencies in performance, and changes in flock management can then be introduced to rectify the unsatisfactory aspects. However, current performance can only be assessed if it is accurately known, and this requires a system of flock recording to be set up.

The flockmaster can, of course, set up his own recording system. Alternatively, he or she will derive considerable benefits from joining a recording scheme such as those on offer from the MLC and ADAS. As an example, the ADAS 'Sheep Monitor' and 'Sheep Master' schemes monitor the flock to calculate physical and financial performance and include five advisory visits each year to discuss the following.

- Preliminary appraisal of the flock.
- Mating management.
- Winter feeding.
- Lambing and grassland management.
- Lamb marketing.

Flock records

The most important records to be kept (in or out of an official recording scheme) are the following.

- Diary of events: Animal health treatments, body condition scoring and pregnancy scanning results, dates of tupping, lambing and sales.
- Livestock numbers: (a) Opening and closing valuations. (b) Ewes put to the tup. (c) Losses (with causes) and sales of ewes and rams. (d) Purchases of breeding stock and store lambs.
- Lambing records: (a) Lambs born alive and dead. (b) Lamb losses, indicating the causes. (c) Sales of finished lambs (live or dead weight) with dates and weights.
- Grazing and conservation records.
- Costs of production: (a) *Feed usage.* Purchased and home-grown forage, roots and concentrates. (b) *Forage costs.* To include seeds, fertilisers, sprays and contractors' charges. (c) *Other variable costs.* Including contract and casual labour, away wintering, vets' fees and medicines, haulage and transport, dips and bedding.
- Wool sales, subsidy and bonuses.

Flock Physical Performance

Honesty is the only policy which will lead to improvement. Armed with carefully compiled records, the flockmaster can check out his or her own results by comparing them with published standard figures. These must be realistic, and examples appropriate to spring (March/April) lambing from 414 flocks recorded by the MLC and from Rosemaund EHF are given in Table 13.1. Similar figures from the early lambing flock at Rosemaund EHF and from 47 early lambing flocks recorded by MLC are given in Table 13.2. These flocks lamb down from December to February.

A comparison of flock performance with the standards in Tables 13.1 and 13.2 will indicate clearly any of the following weaknesses.

- Ewe deaths too high.
- Too many barren ewes; this is less important in the early lambing flock as the barreners can be identified by pregnancy scanning and be re-mated in October.
- Low lamb survival rate.
- Proportion of lambs sold finished is too low; this, of course, does not apply where the objective is the production of store lambs.

Table 13.1 Physical performance: spring lambing lowland flocks

	MLC average 1987	*MLC top-third 1987*	*Rosemaund EHF 1987*
Ewes put to tup	100	100	100
Ewes died	5	5	1
Ewes empty	6	5	1
Ewes lambed	92	93	98
Lambs born alive	160	165	200
Lambs sold/retained	145	153	180
Average carcass weight (kg)	17.1	17.4	17.5

Sources: MLC and Rosemaund EHF.

Table 13.2 Physical performance: early lambing lowland flocks

	MLC average 1987	*MLC top-third 1987*	*Rosemaund EHF 1987*
Ewes put to tup	100	100	100
Ewes died	5	5	1
Ewes empty	6	3	2
Ewes lambed	93	96	97
Lambs born alive	152	167	160
Lambs sold/retained	137	155	152
Average carcass weight (kg)	17.6	17.0	16.5

Sources: MLC and Rosemaund EHF.

Steps taken to improve any of these aspects of performance and also the lamb gradings will benefit the financial outcome.

Flock Financial Performance

The gross margin is probably the best yardstick for the comparison of the performance of sheep flocks operating similar systems and the performance of similar farming enterprises.

The gross margin system

The following is an example layout.

- *Output*:
  value of all lambs produced (finished, store or retained as flock replacements);
  cull ewes and rams;
  ewe premium;
  wool.

  less ewe and ram replacement costs
  Total output.
- *Variable costs*:
  concentrates (for ewes and lambs);
  veterinary surgeon and medicines;
  other expenses (e.g. contract and casual labour, marketing);
  forage costs.
  Total variable costs.
- *Gross margin* = *output* less *variable costs*

Variable costs are allocated readily to an enterprise and vary in direct proportion to the size of the enterprise.

The gross margin does not directly reflect profit. To arrive at this we must subtract from the farm total of gross margins the fixed or overhead costs of the farm. More of this later.

The financial comparison of farm enterprises

It has already been pointed out that the gross margin technique is valid for comparing similar enterprises; this is because dissimilar enterprises may have widely differing effects on the farm's fixed costs and therefore the ultimate profit may show no correlation with the gross margin figure. However, it is interesting to compare sheep with other livestock and crop enterprises in terms of gross margin per hectare. By this means we can form a judgement on the level of farm fixed costs which are sustainable by a ewe flock as against, for example, the growing of winter wheat or the running of a dairy herd or a beef enterprise (Table 13.3).

In terms of gross margin per hectare, lowland ewe flocks are likely to be lower than cash root crops, dairy herds and semi-intensive beef units. However, the performance of lowland ewe flocks may be better than that of oil seed rape, winter barley and single-suckled beef cows.

Table 13.3 Gross margin budgets

Enterprise	*Gross margin per hectare 1988* (£)	*Gross margin per hectare 1989* (£)
Potatoes (main crop)	1770	1750
Friesian dairy herd (autumn calving)	1089	1250
Beef production from grass silage	1004	1250
Sugar beet (mineral soils)	917	900
18 month beef production (autumn born)	665	700
Winter wheat (first crop after break)	591	565
Spring (Easter) lamb production (indoor finish)	540	590
Summer lamb production (lowland)	518	500
Winter oil seed rape	451	480
Winter barley (on good soils)	420	470
Single-suckled beef (lowland, autumn calving)	325	—

Source: ADAS.

The financial comparison of sheep enterprises

The present performance of the best lowland flocks can be considerably better than the gross margins of £500–590/ha given as reasonable targets in Table 13.3. This is illustrated by the MLC 1987 results from 414 spring lambing and 47 early lambing flocks which show both average gross margin performance, and that of the top-third flocks. These are given in Tables 13.4 and 13.5.

The higher stocking rate achieved by the top-third flocks was the most important reason for their superiority. These flocks also produced more lambs per ewe (Table 13.1), sold lambs at a higher average price, had a lower flock replacement cost and a slightly lower cost of feed and forage.

The superior performance of the top-third early lambing flocks resulted mainly from their higher stocking rate and their higher number of lambs reared per ewe (Table 13.2).

How can the flock gross margin be optimised?

The answer becomes clear from a study of the performance of average and top-third spring lambing flocks as recorded by the MLC and shown in Table 13.4.

Table 13.4 Results from 414 lowland spring lambing flocks selling most of their lambs off grass in summer and autumn 1987

		Average (£/ewe)		*Top third* (£/ewe)
Output				
lamb sales[1]	57.38		61.71	
wool sales	3.06		3.10	
ewe premium and LDA subsidy	4.82		4.85	
gross returns		65.26		69.66
Less flock replacements		9.79		7.92
Output		55.47		61.74
Variable costs				
ewe concentrates[2]	7.54		7.01	
lamb concentrates[3]	1.59		1.91	
purchased forage	0.75		0.80	
fertiliser	4.26		3.98	
other forage costs	0.70		0.46	
rented grass keep	0.42		0.52	
total feed and forage	15.26		14.68	
vet and medicine	3.80		3.58	
miscellaneous and transport	1.44		1.34	
Total variable costs		20.50		19.60
Gross margin (output − variable costs)		34.97		42.14
Stocking rate/ha (overall grass and forage)		12.50		15.60
Gross margin/ha		437.00		657.00

	Average	Top third
[1] Average return/lamb (£)	38.00	39.06
Estimated return/kg lamb carcass (£)	2.22	2.25
[2] Ewe concentrate cost/t (£)	135.00	128.00
Ewe concentrates/ewe (kg)	56.00	55.00
[3] Lamb concentrate cost/t (£)	159.00	159.00
Lamb concentrates/ewe (kg)	10.00	12.00

Source: MLC *Sheep Yearbook 1988*.

It is noticeable that the variable costs of the top-third flocks are similar to those of the average flocks. The costs of flock replacement, forage, veterinary services and medicines are practically identical. The better-performing flocks spent a little less on ewe and lamb concentrates (£8.92 vs. £9.13/ewe). There appears to be little scope for increasing margins by means of reducing variable costs although, of course, an eye must always be kept on cost control and, in particular, avoidance of wastage of feed and medicines.

Table 13.5 Results from 47 early lambing flocks for 1987

		Average (£/ewe)		*Top-third* (£/ewe)
Output				
lamb sales[1]	69.56		75.07	
wool sales	2.81		2.76	
ewe premium	4.94		4.74	
gross returns		77.31		82.57
Less flock replacements		9.14		7.39
Output		68.17		75.18
Variable costs				
ewe concentrates[2]	9.68		8.81	
lamb concentrates[3]	12.31		14.22	
purchased forage	1.44		1.40	
fertiliser	3.54		2.68	
other forage costs	0.47		0.31	
rented grass keep	0.05		0.11	
total feed and forage	27.49		27.53	
vet and medicine	3.93		3.18	
miscellaneous and transport	1.89		1.76	
Total variable costs		33.31		32.47
Gross margin (output − variable costs)		34.86		42.71
Stocking rate/ha (overall grass and forage)		15.20		18.60
Gross margin/ha		530.00		794.00
[1] Average return/lamb (£)		48.31		47.82
Estimated return/kg lamb carcass (£)		2.74		2.82
[2] Ewe concentrate cost/t (£)		136.00		133.00
Ewe concentrates/ewe (kg)		71.00		66.00
[3] Lamb concentrate cost/t (£)		156.00		155.00
Lamb concentrates/ewe (kg)		79.00		92.00

Source: MLC *Sheep Yearbook 1988*.

The top-third flocks achieve their superiority by their higher output in terms of lamb sales per ewe (£61.71 vs. £57.38) and lamb sales per hectare (£963 vs. £717), the latter influenced by a higher stocking rate of 15.6 as against 12.5 ewes/ha for the average flocks. It is clear that the most successful flockmasters are a little better at controlling their costs and a lot better at maximising their output.

High output is the key to the financial success of the flock, and it means that the overheads or fixed costs (including those of labour) are spread over more units of production.

Raising flock output

All possible efforts should be made to raise output per ewe, not least because this has a considerable influence on output per hectare.

Financial output per ewe

The main component is lamb sales either finished or store. Sales of wool together with those of cull ewes and rams are normally only about 10 per cent of total sales.

The return from lamb sales depends upon the number sold per ewe and their unit value. Prolific crossbred ewes now freely available are capable of a lambing percentage per ewe put to the tup of 200, and a lamb sale percentage of 180. An improvement in lambing percentage and a reduction in lamb mortality need not be expensive. The above figures can be achieved in the lowland flock without expensive hormonal aids or high levels of supplementary compound feeding; they depend on good well-managed grassland, and the best sheep husbandry and disease prevention techniques. Unit value of the lambs sold may be boosted by selling a high percentage finished as opposed to store, by selling during periods of high lamb prices, and by marketing lambs of the weight and quality the buyers prefer.

Output per hectare

Lamb sales per hectare are the product of sales per ewe and the stocking rate at grass. Over the last 15 years at Rosemaund EHF, the ewes have been stocked at 15/ha on perennial ryegrass/white clover leys receiving 120 kg N/ha. The grass has provided all the summer grazing and also the conserved bulk feeds for the winter. In recent years, the winter feeding of the ewes on straw instead of grass silage has raised the stocking rate dramatically to reach a level of 30 ewes/ha from 1986 onwards.

What is the best combination of lamb sales per ewe and stocking rate?

At a given stocking rate, all increases in lamb sales per ewe will increase the output per hectare; this is illustrated in Table 13.6, assuming a lamb sale value of £35 and a stocking rate of 16 ewes and their lambs/ha. Similarly, at a given level of lamb sales per ewe all increases in stocking rate increase the output per hectare. Table 13.7 assumes a lamb sale value of £35 and lamb sales of 1.6/ewe.

Table 13.6 The effect of numbers of lambs sold per ewe on output per hectare

Number of lambs sold per ewe	*Number of lambs sold per hectare*	*Lamb output per hectare (£)*
1.50	24.0	840
1.60	25.6	896
1.70	27.2	952
1.80	28.8	1008
1.90	30.4	1064
2.00	32.0	1120

Source: Rosemaund EHF.

Table 13.7 The effect of stocking rate on output per hectare

Stocking rate of ewes per hectare	*Number of lambs sold per hectare*	*Lamb output per hectare (£)*
10	16.0	560
12	19.2	672
14	22.4	784
16	25.6	896
18	28.8	1008
20	32.0	1120
22	35.2	1232
24	38.4	1344
26	41.6	1456
28	44.8	1568
30	48.0	1680

Source: Rosemaund EHF.

Lamb sales

Lamb sales per ewe are limited by the potential prolificacy of the animals. This potential may be increased by the purchase of highly prolific ewes or by the use of hormones to improve prolificacy or to increase the frequency of lambing. However, these measures mean higher costs including those of the additional labour needed to keep undersized multiple-birth lambs alive, and consequently most flockmasters show little enthusiasm for ewes of very high prolificacy. Sales of 1.8 lambs/ewe from a flock with a lambing percentage of around 200 represent a good target without going over the top.

Stocking rate

We can then seek to combine the above level of lamb sales per ewe with the highest possible stocking rate. Until recently, it was accepted that increased stocking density meant decreased individual lamb performance. However, several developments have allowed us to intensify the stocking rate and yet maintain ewe performance and lamb-weight gains at a high level. These developments are: better adjustment of grass grazing height, the supplementary feeding of ewes and lambs at grass, a change from hay to silage making and, above all, more efficient control of worm parasites. With these aids it is clear that a stocking rate of 15 ewes/ha of grass on a year-round basis can be maintained comfortably, and 27 lambs sold/ha/year.

The more recent development of basing the winter feeding of the ewe flock on treated or untreated straw means that grassland conservation in the form of silage or hay is unnecessary and the consequent increase in stocking density at grass can have big effects on both the physical and financial performance of the flock. This is illustrated in Tables 13.8–10 giving the results from experiments at Rosemaund EHF carried out over a 4-year period.

The increases in stocking density made possible by the abandonment of grass conservation and the winter feeding of straw led to increased lamb output per hectare.

The creep feeding of lambs ensured their early sale, so easing the cash flow. Set stocking at 22 ewes and their lambs/ha with no creep feeding produced a large number of store lambs.

The winter feeding of straw facilitated a very high stocking rate in the Rosemaund EHF experiments, and this had a favourable effect on the

Table 13.8 Total output of lamb (kg/ha)

Stocking rate	*22 ewes/ha reduced in June to 15 ewes/ha*	*22 ewes/ha*	*22 ewes/ha plus creep feed*	*30 ewes/ha plus creep feed*
	(Grass conserved for winter silage feeding)	*(Straw-fed in winter)*	*(Straw-fed in winter)*	*(Straw-fed in winter)*
1984	945	1393	1358	—
1985	897	1256	1343	—
1986	895	1203	1313	1883
1987	—	—	1400	1909

Source: Rosemaund EHF.

Table 13.9 Percentage of lambs sold finished and as stores

Stocking rate	*22 ewes/ha reduced in June to 15 ewes/ha (average of 3 years)*	*22 ewes/ha (average of 3 years)*	*22 ewes/ha plus creep feed (average of 4 years)*	*30 ewes/ha plus creep feed (average of 2 years)*
Sold finished	38	20	100	93
Sold store	62	80	0	7

Source: Rosemaund EHF.

Table 13.10 Gross margins (£/ha)

Stocking rate	*22 ewes/ha reduced in June to 15 ewes/ha*	*22 ewes/ha*	*22 ewes/ha plus creep feed*	*30 ewes/ha plus creep feed*
1984	704	881	894	—
1985	608	813	750	—
1986	667	770	842	928
1987	—	—	729	1008

Source: Rosemaund EHF.

gross margin per hectare. However, the lowland straw feeding flocks recorded by the MLC have not shown the improvement in performance consistently achieved at Rosemaund since 1984.

The importance of fixed costs

The *management and investment income* is the margin available to cover the return on tenant's capital and the reward to management. It is worked out on the basis of the 'profit' a farmer would make after 'paying' a wage for the physical work he and his family do on the farm, and paying a rent for the farm, even if owned, but before allowing for any interest charges that may be due on borrowed money.

Management and investment income = total farm gross margin *less* fixed costs.

The *fixed or overhead costs* are those which cannot easily be allocated to an enterprise and change little with small adjustments to the farm system. They include regular labour, machinery repairs and depreciation, fuel, electricity, insurance and office expenses.

Fixed costs are characterised by their wide variation from farm to farm. However, on predominantly cattle and sheep farms recorded in 1986 by the University of Reading Department of Agricultural Economics and Management they averaged £412.6/ha.

The scope for reduction of fixed costs

It is difficult to see how changes in the management of the sheep flock can significantly reduce the level of the farm's fixed costs. The only sector of these susceptible to substantial reduction is labour. Modern thinking is that the trained shepherd should be able to look after a flock of at least 800 lowland ewes, and some would say that this figure should be 1000. There are, however, two important provisions. Firstly, he must have assistance during the peak work periods such as lambing, shearing and dipping and this is facilitated by avoiding clashes with peak labour periods in other enterprises; winter shearing, for example, helps here. Secondly, he must be provided with a well-designed labour-saving sheep-handling set-up (to include a sheep cradle) and easy access to well-fenced grazing fields.

Capital considerations

The sheep enterprise benefits from having a relatively low fixed-capital requirement, which is why sheep numbers can expand and contract quite rapidly in response to trends in profitability. The handling unit and fences have already been mentioned, and the capital implications of the use of winter housing have been discussed in Chapter 3. However, the requirement of working capital per hectare of land utilised by sheep is substantial, especially where the stocking density is high in the more intensive units. The saving grace is that most of it is invested in the flock itself, and therefore is readily realisable.

The *working capital* is the cost of the breeding sheep (ewes and rams) plus half the variable costs incurred, e.g. the working capital per hectare might be as shown in Table 13.11.

Return on working capital

The gross margin can be related to the capital requirement to give a return on working capital. This return may then be compared with the return on working capital invested in alternative enterprises,

Table 13.11 Working capital requirement

	(£)
15 ewes at £80	1200
Half a ram at £400	200
Half total variable costs at £300	150
Working capital per hectare	1550

Table 13.12 Gross margin as return on working capital

Main winter feed	*Gross margin per hectare* (£)	*Working capital per hectare* (£)	*Gross margin per £100 capital* (£)
Silage – 15 ewes/ha	660	1550	42.6
Straw – 22 ewes/ha	804	2270	35.4
Straw – 30 ewes/ha	968	3100	31.2

Source: Rosemaund EHF.

Table 13.13 Gross margin less finance cost

Main winter feed	*Gross margin per hectare* (£)	*Working capital per hectare* (£)	*Finance cost at 12%* (£)	*Gross margin less finance cost per hectare* (£)
Silage – 15 ewes/ha	660	1550	186	474
Straw – 22 ewes/ha	804	2270	272	532
Straw – 30 ewes/ha	968	3100	372	596

Source: Rosemaund EHF.

and a conclusion reached on the viability of the sheep enterprise. (Table 13.12).

Gross margin less finance cost

A further measure which can be used to compare performance with that of other enterprises is the gross margin *less* the finance cost; this indicates the margin left to cover other overheads, and is shown in Table 13.13.

Flock Targets

Flock targets are best expressed in terms of physical performance, and the following are suggested.

- *Lowland flocks where grassland provides the summer keep and the conserved winter feed: 900 kg lamb live-weight/ha.*

  This may be achieved by selling 1.7 lambs/ewe at a year-round stocking rate of 15 ewes/ha and an average sale weight of 35 kg/lamb.

  Weight of lamb sold/ewe = 1.7 × 35 = 60 kg
  Weight of lamb sold/ha = 60 × 15 = 900 kg
- *Lowland flocks at grass in the summer and fed by-product straw in the winter: 1300 kg lamb live-weight/ha.*

  Where straw replaces hay or silage as the bulk winter feed the stocking rate at grass may be increased to at least 22 ewes/ha.

  Weight of lamb sold/ewe = 1.7 × 35 = 60 kg
  Weight of lamb sold/ha = 60 × 22 = 1320 kg

The performance achieved at Rosemaund EHF over the last 4 years is similar to the above targets. This can be seen in Table 13.14 which also shows the corresponding financial performance in terms of gross margin per hectare.

It should be noted that the gross margin does not rise in proportion to the lamb output because the costs of supplementary compound feeds are much higher where straw replaces silage as the winter bulk feed. Please note also that these results were obtained on high-quality perennial ryegrass/white clover leys grown on fertile water-retentive silty soils in Herefordshire with adequate rainfall (average

Table 13.14 Flock performance at Rosemaund EHF 1984–87

Main winter feed	*Stocking rate per hectare*	*Lamb sold per hectare* (kg)	*Gross margin per hectare* (£)
Silage (average of 3 years)	15	912	660
Straw (average of 4 years)	22	1353	804
Straw (average of 2 years)	30	1896	968

Source: Rosemaund EHF.

665 mm/year). In less 'grassy' areas this level of performance could not be achieved. However, in all areas the flock output and the margin will be optimised by attention to the following ten key aspects of management.

A Blueprint for Success

- *The grassland.* A mixed sward of grasses and clover is preferred. It allows the ewe flock to be densely stocked and the lambs to make good weight gains. On such swards limited applications of nitrogenous fertiliser totalling 120 kg/ha ensure adequate forage growth, the persistence of the clover and low grass-growing costs. Outside the 'grassy' areas, or where clover is lacking in the sward, the nitrogenous fertiliser requirement is increased to around 180 kg/ha, and to over 200 kg/ha on thin soils.
- *The sheep.* Several breeds and crosses of ewe have the potential for 200 per cent lambing. The choice should be made from among these on the basis of ready market availability and reasonable cost. It is unwise to purchase ewes of very high prolificacy (such as Finnish Landrace crosses) unless there is a willingness to rear numerous triplet and quadruplet lambs.

  Rams of good conformation and growth rate should be purchased, favouring the earlier maturing breeds to produce standard-weight lambs and the later maturing breeds to produce heavy lambs or lambs for the store market.
- *Flock health.* For reasons of animal welfare and flock productivity it is vitally important to set up a year-round health-care programme. This should be formulated after detailed discussions with your veterinary surgeon.
- *Sheep housing and handling.* The winter housing of ewes may not be necessary on well-drained land. However, on land which poaches and for highly prolific or early lambing flocks it greatly benefits the sheep and the shepherd. A well-designed handling unit is essential if the flock is to be dipped, treated and sorted on time with minimal labour input.
- *Output per ewe.* Financial success depends upon producing a heavy lamb crop and achieving a high lamb sale price. It results from:
  (a) *High conception rate.* This is dependent on getting the ewes into the right body condition prior to tupping and maintaining their condition through early pregnancy. Regular body scoring is essential during this period. The use of Fecundin could be considered.

(b) *High lambing percentage.* Accurate winter feeding means lower ewe mortality and lower foetal losses. It is aided by winter housing, separate penning for pregnant ewes according to expected lambing date, body condition scoring and pregnancy scanning where appropriate.

(c) *Low lamb mortality.* Achieved by adequate feeding of the pregnant ewes (fewer small lambs) and by plenty of supervision and the best hygiene at lambing time. Again, winter housing helps, as can pregnancy scanning and winter shearing, particularly in prolific flocks.

(d) *High lamb value.* This means selling the type of lambs the market wants at the right time. With spring lambing flocks every effort should be made to sell finished lambs by the end of June. Early lambing in December/January produces high-value Easter lambs and spreads the workload, but costs are high. Late lambing in May and store lamb finishing also allow the sale of high-value lambs out of season.

- *Output per hectare.* High ewe-stocking rates can be added to high output per ewe by careful management of good grass, and in particular by controlling parasitic worms. The replacement of hay and silage by straw as the main winter ewe feed can increase the stocking density by 50 per cent.
- *Winter feeding.* On the all-grass farm, hay may be the main winter feed but silage is to be preferred where the flock is housed. On mixed farms, arable by-products and root crops come into their own. Straw, correctly supplemented, is a completely satisfactory ewe feed. Its use does not reduce total feed costs but does allow an increased stocking rate at grass. It also means an easier life-style free of the stresses of hay and silage making!
- *Lamb growth rate.* A short dense clovery sward must be maintained and parasitic worms controlled by the clean-grazing technique or by carefully timed anthelmintic drenches. Early lamb growth may be improved by supplementary feeding of ewes with twins during the first 6 weeks at grass and later lamb growth increased by offering creep feed. Experimental evidence favours the 12 weeks weaning of lambs on to parasite-clean grass.
- *Marketing.* During the selling period, regular weighing and handling of lambs is essential in order to meet market preferences in terms of weight, fat, conformation and uniformity. Carcass damage must be avoided, particularly bruising, abscesses and dirty fleeces.
- *Stockmanship.* The most important requirement has been left to the last. The possession of the finest grassland, sheep and

handling facilities will be brought to nought without high-quality stockmanship.

It is the combination of skilled shepherding and the proven technical advances of recent years which can best ensure the health and performance of the flock. Lamb sales of 1300 kg/ha can be achieved, giving a gross margin of £800/ha. The ewe flock is then no longer a difficult-to-justify subsidiary unit, but an enterprise which can use land more profitably than a crop of cereals.

Appendix A

Nutritive Values of Selected Feeds

	Dry matter (%)	*Metabolisable energy* (MJ/kg DM)	*Digestible crude protein* (g/kg DM)	*Crude protein degradability*
Fresh herbage				
high-quality	20	13.1	125	M
medium-quality	20	11.6	105	M
low-quality	23	9.8	60	M
Grass hay				
high-quality	85	10.3	77	H
medium-quality	85	8.7	70	H
low-quality	85	8.1	55	H
Silage				
grass of high quality	25	11.8	125	H
grass of medium quality	25	10.6	105	H
grass of low quality	25	9.2	100	H
maize	25	11.5	50	H
Roots				
swedes	11	14.0	65	H
turnips	10	12.7	70	H
fodder beet	18	11.9	35	H
Straws				
wheat – untreated	86	6.1	10	—
barley – untreated	86	6.5	16	—
barley – ammonia treated	86	7.4	20	—
Cereal grains				
barley	86	12.8	90	H
wheat	86	13.6	118	H
oats	86	12.0	85	M
Concentrates				
dried molassed sugar-beet pulp	86	12.5	80	M
extracted soya bean meal	90	13.4	455	M
white fish meal	90	14.2	690	L

Source: ADAS Booklet P2087 *Nutrient Allowances for Cattle and Sheep*. This booklet also gives the nutritive value of other foods.
NB: Rumen degradability of protein estimated as high (H), medium (M) or low (L).

Appendix B

The Shepherd's Calendar

March lambing flock

July	Dip the ewes and remaining lambs. Condition-score the ewes.
August	Apply nitrogen to grass at 40 kg/ha. Apply phosphorus and potassium fertiliser to grass. Purchase replacement ewes and rams. Dose these against worms and fluke and vaccinate them fully against clostridial diseases and pasteurellosis. Vaccinate any remaining lambs against clostridial diseases.
Early September	Condition-score ewes and check health including feet. Check rams' condition and health including feet.
Late September	Complete sale of fit lambs. Put remaining lambs on store feeding rations and buy additional store lambs. Flush the ewes. Vaccinate against enzootic abortion if necessary.
Early October	Dip ewes, rams and store lambs. Check and worm the rams.
20 October	Turn raddled rams in with the ewes.
4 November	Change raddle colour.
20 November	Change raddle colour.

Late November	Vaccinate against foot rot if necessary. Have samples of hay/silage analysed. Blood test to check selenium status if necessary.
December/January	Condition-score the ewes. House the flock, penning according to expected lambing date and condition score. Worm the ewes. Check and trim ewes' feet. Inject booster vaccine against foot rot. Winter shearing of ewes.
Early January	Pregnancy scanning of ewes.
January/February/March	Run housed ewes through zinc sulphate foot bath every 14 days.
February/March	Spray cobalt sulphate on to sheep pastures in 'pine' areas.
23 February	Inject all ewes with booster vaccine against clostridial diseases and pasteurellosis.
Early March	Check fencing of sheep pastures. Apply nitrogen to grass at 40 kg/ha. Check shepherd's cupboard. Erect the individual lambing pens.
10 March	First lambs will appear.
15 March	Lambing starts in earnest. Dress navels of new-born lambs, castrate and tail. Turn out ewes and lambs after 24–48 hours in lambing pens. Worm ewes at turn out if not done at housing.
3 April	All ewes conceiving to a first mating will have lambed so lambing rate will slow and shepherd can go to bed!
March/April	Feed concentrate containing calcined magnesite to ewes at grass for 6 weeks after lambing.

	Creep-feed densely stocked lambs at grass from turn out.
Late April	Drench lambs against worms and repeat every 4 weeks if not on clean grass. Watch carefully for coccidiosis, and treat immediately if it occurs.
May	Commence marketing of lambs.
May/June	Shearing of ewes (if not winter clipped). Dip the lambs against fly blow if necessary.
Mid June	Apply nitrogen to grass at 40 kg/ha.
July	Wean the lambs, drench and put on clean grass or forage crops. Dry-off the ewes. Cull the ewes.

Early lambing flock

Grassland husbandry and most sheep-health protection measures are as for the spring lambing flock.

Early June	Condition-score the ewes.
Early July	Vaccinate previously unvaccinated sheep fully against clostridial disease and pasteurellosis. Vaccinate against enzootic abortion if necessary. Flush the ewes. Check and worm the rams.
28 July	Insert sponges.
9 August	Remove sponges and inject ewes with PMSG.
11 August	Turn raddled rams in with the ewes.
26 August	Change raddle colour.
Late October	Condition-score the ewes. Blood test to check selenium and copper status if necessary.

Mid December	Inject booster vaccine against clostridial diseases and pasteurellosis.
December	House the flock.
2 January	Induce the ewes to lamb.
4 January	Commencement of lambing.
14 January	Start creep-feeding of lambs.
Mid February	Wean the lambs.
Early March	Commence marketing of lambs.

Appendix C

Body Condition-Scoring*

Condition-scoring is a standard handling technique to assess how much muscle and fat breeding ewes are carrying at a particular time. It is a vital tool in ewe management. Detailed records of the condition of ewes at mating and of their subsequent prolificacy have proved without doubt that thin and fat ewes do not perform to their full ability.

Body condition-scoring should be a part of the shepherd's routine at three times of the year as follows.

- Eight weeks before tupping. Target score 3.5 at mating. The time lapse of 8 weeks allows remedial action to be taken so that all ewes achieve the target score by the time the tups are turned in.
- In mid pregnancy. Target score 3.0.
- In late pregnancy. Target score 3.0–3.5 for housed ewes and 3.5–4.0 for out-wintered ewes.

How to condition-score ewes

Body condition is assessed by handling the ewe over and round the backbone, in the area of the loin behind the last rib and above the kidneys.

Using the finger tips, firstly feel the degree of sharpness or roundness of the lumbar vertebrae as in Figure 1, position A. Secondly, feel and assess the prominence and degree of cover over the horizontal processes (position B). Then assess by feel the amount of muscle and fat under the horizontal processes, by the ease with which the fingers pass under the ends of these bones. Finally, assess the eye muscle and its fat cover, by pressing the fingers into the area between the vertical and horizontal processes.

Taking these assessments into account, it is usual to score all ewes

* This appendix is based on information in the MLC Sheep Improvement Services Booklet *Body Condition Scoring of Ewes* and ADAS Leaflet 787, *Condition Scoring of Ewes.*

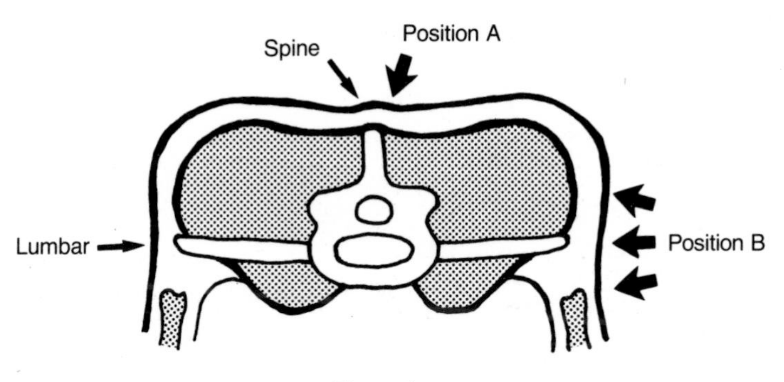

Figure 1

on a scale of 0–5, using half-scores as intermediate points along the scale.

- *Score 0 (emaciated).* This is seldom used as it only applies to ewes which are extremely emaciated and on the point of death. It is not possible to feel any muscle or fatty tissue between skin and bone.
- *Score 1 (very lean).* The vertical (spine) and horizontal (lumbar) processes are prominent and sharp. The fingers can be pushed easily below the horizontals and each process can be felt. The loin muscle is thin and with no fat cover (Figure 2).

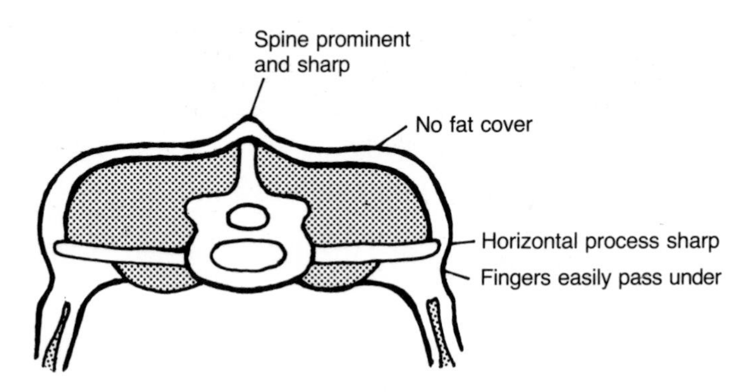

Figure 2

- *Score 2 (lean).* The vertical processes are prominent but smooth, individual processes being felt only as corrugations. The horizontal processes are smooth and rounded, but it is still possible to press the fingers under. The loin muscle is of moderate depth but with little fat cover (Figure 3).

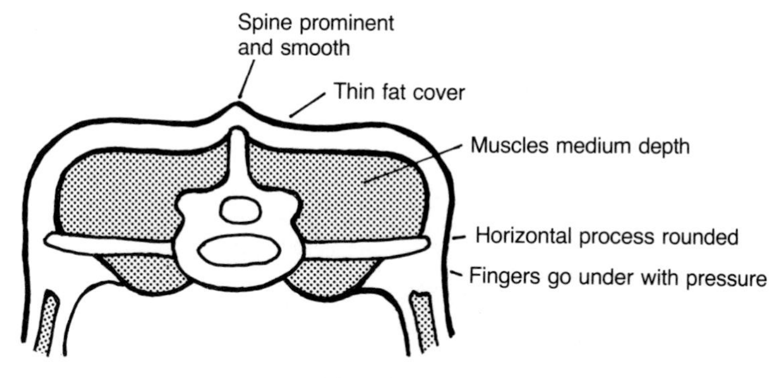

Figure 3

- *Score 3 (good condition)*. The vertical processes are smooth and rounded; the bone is only felt with pressure. The horizontal processes are also smooth and well covered; hard pressure with the fingers is needed to find the ends. The loin muscle is full, with a moderate fat cover (Figure 4).

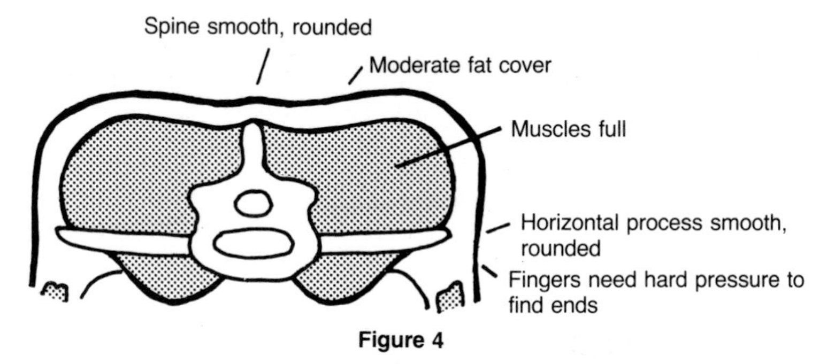

Figure 4

- *Score 4 (fat)*. The vertical processes are only detectable as a line; the ends of the horizontal processes cannot be felt. The loin muscles are full and have a thick covering of fat (Figure 5).

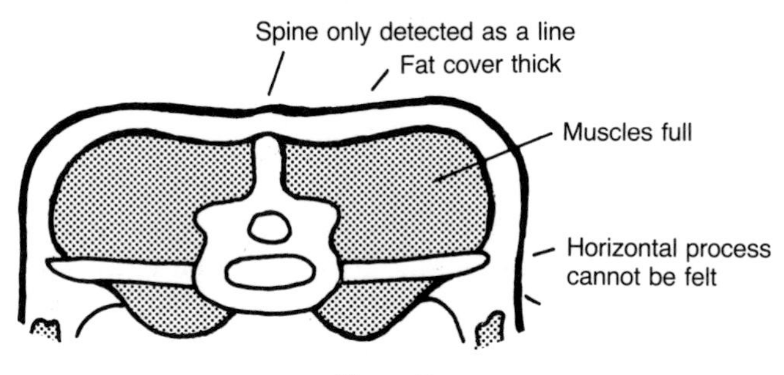

Figure 5

- *Score 5 (very fat).* The vertical processes cannot be detected even with pressure; there is a dimple in the fat layers where the processes should be. The horizontal processes cannot be detected. The loin muscles are very full and covered with very thick fat (Figure 6).

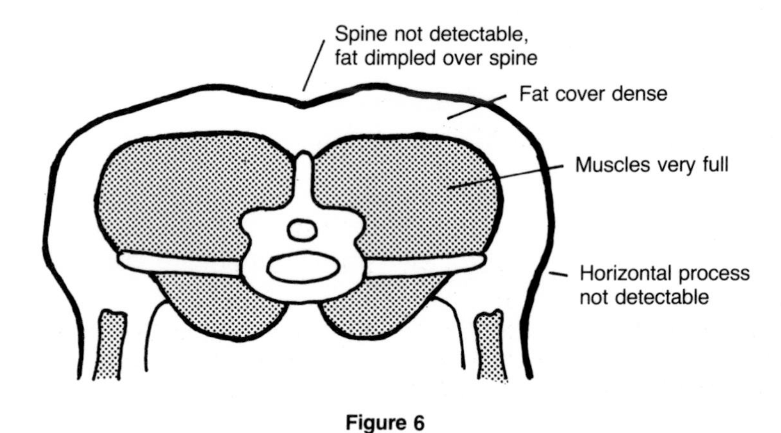

Figure 6

Figures 1–6 from MLC

Appendix D

Conversion Factors: Metrication and Abbreviations

Length

cm (centimetres) to in	× 0.394
mm (millimetres) to in	× 0.0394
m (metres) to ft	× 3.29
m (metres) to yd	× 1.09
km (kilometres) to miles	× 0.621

Weight

g (grams) to oz	× 0.0353
g (grams) to lb	× 0.0022
kg (kilograms) to lb	× 2.20
kg (kilograms) to cwt	× 0.0197
t (tonnes) to tons	× 0.984

Area

m^2 (sq metres) to sq ft	× 10.8
m^2 (sq metres) to sq yds	× 1.20
ha (hectares) to acres	× 2.47

Temperature

(°C × 1.8) + 32 = °F

Some double conversions

kg/hectare	=	fertiliser units/acre	× 1.25
t/ha	=	cwt/acre	× 0.125
kg/ha	=	lb/acre	× 1.1
litres/ha	=	pints/acre	× 1.4

Index

FARMING PRESS BOOKS

The following are samples from the wide range of agricultural and veterinary books published by Farming Press. For more information or for a free illustrated book list please contact:

Farming Press Books, 4 Friars Courtyard
30–32 Princes Street, Ipswich IP1 1RJ, United Kingdom
Telephone (0473) 43011

The TV Vet Sheep Book

EDDIE STRAITON

A pictorial guide to all the common sheep ailments. The concise text includes a major section on lambing.

Profitable Sheep Farming

M. McG. COOPER AND R. J. THOMAS

Chapters on production, breeds, nutrition, management, store lamb feeding, ewe selection and recording, profitability and sheep ailments.

Intensive Sheep Management

HENRY FELL

An instructive practical account of lowland sheep farming based on the experience of a leading farmer and breeder.

A Way of Life: Sheepdog Training, Handling and Trialling

H. GLYN JONES AND BARBARA COLLINS

A complete guide to sheepdog work and trialling, in which Glyn Jones' life is presented as an integral part of his tested and proven methods.

The Blue Riband of the Heather

E. B. CARPENTER

A pictorial cavalcade of International Sheep Dog Society Supreme Champions from 1906, including information on pedigrees, awards and distinguished families of handlers.

Farming Press also publish four monthly magazines: *Livestock Farming*, *Arable Farming*, *Dairy Farmer* and *Pig Farming*. For a specimen copy of any of these magazines, please contact Farming Press at the address above.